Reginald Heber Howe, Edward Sturtevant

The birds of Rhode Island

Reginald Heber Howe, Edward Sturtevant

The birds of Rhode Island

ISBN/EAN: 9783743322202

Manufactured in Europe, USA, Canada, Australia, Japa

Cover: Foto ©berggeist007 / pixelio.de

Manufactured and distributed by brebook publishing software (www.brebook.com)

Reginald Heber Howe, Edward Sturtevant

The birds of Rhode Island

PURGATORY.
WHERE BARN SWALLOWS BREED.

THE
BIRDS OF RHODE ISLAND.

BY

REGINALD HEBER HOWE, JR.,

Member of the Nuttall Ornithological Club.

AND

EDWARD STURTEVANT, S. B.,

Instructor of Natural Sciences at Saint George's School, Newport.

Members of the American Ornithologists' Union.

ILLUSTRATED.

1899.

COPYRIGHT, 1899.

BY R. H. HOWE, JR., AND E. STURTEVANT.

CONTENTS.

PREFACE. 5

PART I.

REVIEW OF FORMER PUBLICATIONS ON RHODE ISLAND
    BIRDS, AND OF STATE COLLECTIONS   7

MIGRATION, WITH LIST OF BREEDING BIRDS   10

CORMORANT ROCK   17

NOTE   22

PART II.

ANNOTATED LIST .   25

EXTIRPATED SPECIES   88

HYPOTHETICAL LIST . . 89

BIBLIOGRAPHY   91

ERRATA, ADDITIONS, ETC. . . 102

INDEX, SCIENTIFIC NAMES   103

    VERNACULAR AND LOCAL NAMES   107

ILLUSTRATIONS.

PURGATORY, Middletown: — Frontispiece.

MOUNT HOPE ON NARRAGANSETT BAY . . . 11

CORMORANT ROCK and Tern's Nest . . . 17

AN OSPREY'S NEST, Bristol 25

A BANK SWALLOW COLONY and Section of Single Nest,
    Sachuest Point 89

ROSE-BREASTED GROSBEAK'S NEST . . . 91

PREFACE.

THIS volume on the Birds of Rhode Island, containing the first complete list of the birds of the State, is offered with the hope that it will lead to the further study of ornithology within Rhode Island, and that it will give a basis on which to build such work.

The authors here wish to express their sincere thanks to Lieut. Wirt Robinson, 4th U. S. Artillery, West Point, New York, Col. J. H. Powel of Newport, Dr. Wm. C. Rives of New York, Messrs. H. S. Hathaway of Cranston, F. T. Jencks of Drownville, J. M. Southwick, Newton Dexter, F. E. Newbury, C. H. Smith, E. H. Armstrong of Providence, George W. Field of Kingston, Owen Durfee of Fall River, Mass., Glover M. Allen of Intervale, New Hampshire, Walter Faxon of Arlington, Mass., H. W. H. Powel, Le Roy King of Newport, William Brewster, Walter Deane of Cambridge, Mass., Miss Louisa Sturtevant of Newport, and many others.

REGINALD HEBER HOWE, JR.
EDWARD STURTEVANT.

RHODE ISLAND,
October, 1899.

PART I.

REVIEW OF FORMER PUBLICATIONS ON RHODE ISLAND BIRDS, AND OF STATE COLLECTIONS.

RHODE ISLAND, though adjoining Massachusetts, a State whose avi-fauna has had long and careful study, is remarkable in that it has had but few ornithologists, and but little ever published in regard to its birds.

The first publication that appeared on the birds for any portion of the State was in 1884 when Dr. William C. Rives, M. A., published a short paper, entitled "The Birds of Newport," in the Proceedings of the Newport Natural History Society, 1883-4, page 28. This paper was one of Dr. Rives' first pieces of ornithological work, and cannot be compared with his admirable work, "A Catalogue of the Birds of the Virginias." The paper is of a purely popular nature, mentioning some ninety-seven species. It is annotated only in a casual way and contains little of scientific value. In the same Proceedings of the Newport Society, page 42, there is a bare "List of Birds Shot Near Newport," by Col. John Hare Powel, unannotated. Col. Powel though an old, well known sportsman in Rhode Island, does not pose as an ornithologist, and does not attempt to establish records upon his own identification. In 1888, Mr. J. M. Southwick published a paper, entitled "Our Birds of Rhode Island," in the Proceedings of the Newport Natural History Society, 1887-8, page 3, of very much the same character as Dr. Rives'. It contains, however, more of value in regard to Rhode Island birds, perhaps the most of any of the few existing papers. The paper speaks of some one hundred and one species in detail, and Mr. Southwick states that he could then report the "capture of at least two-hundred and thirty species." In this same Proceedings there is a paper by Mr. Charles H. Lawton called, "The Water Birds of Newport," page 16. This list, for it hardly amounts to anything more than that, mentions forty-two species. Mr. Howe published in the Bristol Phoenix for April 10th, 1896, an annotated list of the birds of that township, observed by him. This article, called "A List of the Birds of Bristol, R. I., and Adjacent Localities," contains, beside annotations on sixty-seven species, a short account of the topography of Bristol, with a map showing wooded areas. In a book by the same author, entitled "On the Bird's Highway," 1899, there is in the appendix, an unannotated list of the birds observed at Bristol, R. I. A series of articles were published in the Providence Journal entitled "Birds during January" and "February," issue of February 7, 1884, and "Birds during March" and "April," issue of April 7, 1884, by Mr. F. T. Jencks. An article called "Rhode Island Birds," by Mr. J. M. Southwick, was published in the same paper on February 28, 1892, and another article, under the same title, by Mr. N. W. Wright, on September 25, 1898. These, papers cover the special literature on the birds of Rhode Island except for the minor records which have appeared from time to time in The Auk, Nuttall Bulletin, Random Notes, etc., of which there are exceptionally few.[1]

COLLECTIONS.

There are a number of well known ornithological collections within the State containing birds taken in Rhode Island. The best no doubt is the one at Roger Williams Park, Providence, under the curatorship of Mr. J. M. Southwick. This has been assembled by Mr. C. H. Smith of Providence, and presented by him to the city. During the summer of 1899, this collection was installed with other Rhode Island State collections of Zoology and Mineralogy. The collection of some three hundred and fifty specimens is intended to represent all species known to have occurred within the State, and contains many of the rare Rhode Island captures, reference to which are given in the annotated list.

---

[1] See Bibliography.

There are also among the public collections, those of Brown University at Providence, Rhode Island College of Agriculture and Mechanic Arts at Kingston, Franklin Society at Providence, Newport Natural History Society at Newport. Each of these contain rare birds taken within the State. Among the private collections, those of Mr. H. S. Hathaway of Cranston, Mr. Edward Sturtevant of Newport, Mr. J. M. Stainton of Providence, Mr. Harry A. Cash of Pawtucket, Mr. Walter Angell of North Providence, and the oological collection of Mr. B. La Farge of Newport, Mr. F. E. Newbury of Providence, and Mr. C. E. Doe of Providence are deserving of mention. The State of Rhode Island is much richer in collectors than in observers, although this is to be lamented, yet the records we have are most of them substantiated by the existence of specimens.

MIGRATION.

The migration of birds in Rhode Island is of such a peculiar nature that it seems worthy of especial attention.

WATER BIRDS.

The migration of water birds (*Pygopodes, Longipennes, Tubinares, Steganopodes, Anseres,* and *Limicolæ*) along the Rhode Island coast is very much less pronounced than would be supposed. The main line of migration going north and south, seems to be to a great extent, off the coast a number of miles. From Watch Hill to Point Judith and at Sakonnet Point the greatest migration movement is apparent, and yet, even at these points, the most exposed to the ocean of any portion of Rhode Island, save Block Island, the main line of migration still seems to be further seaward. The centre of the migration flight apparently passes just off, and along the ocean coast of Long Island, past Block Island, to Martha's Vineyard and Nantucket waters. Only the edge of this migration, and a smaller migration that passes through Long Island Sound, up Buzzard's Bay and across Cape Cod, brings birds by the coast of Rhode Island. Therefore many species of water birds which are common at Long Island, Martha's Vineyard and Nantucket, are uncommon on the Rhode Island coast. Few birds land on Block Island, and the movement there is so strictly a direct migratory one that it cannot be compared, as a point of observation, with the above named larger islands.

There is but little migratory movement in Narragansett Bay it being chiefly used by species as a locality in which to rest, or in which to remain for certain seasons. There are, however, a few quite marked local migratory movements both along the coast and in the bay. The westward migration of White-winged

Mount Hope on Narragansett Bay.
From "*On the Birds' Highway.*"

Scoters (*Oidemia deglandi*) in May is the most pronounced and interesting of the local outside movements. In The Auk, Vol. VIII, No. 3, page 285, there is a careful account of this migration by Mr. George H. Mackay, from which we here quote. The Scoters wintering in southern Cape Cod waters migrate "to the westward as far as Noank, Connecticut," past Seaconnet, Point Judith and Watch Hill, reaching the north, Mr. Mackay suggests, by "Connecticut River and Lake Champlain or Hudson River routes." This migration lasts from "three to seven days, according to the state of the weather," starting "May 7, which is unusually early; the customary time being from the 12th to the 15th, and the latest the 25th." The flight consists of "apparently all old birds," and in such fine adult plumage and of such large size that the local gunners believe them to be of a different species from the other less mature White-winged Scoters seen throughout the winter, and have named them May White-wings or Great May White-wings for this reason. Mr. Newton Dexter writes of this flight, "In May they gather in millions, I might say, about Vineyard Sound, and farther east. About May 17th if the conditions are right, fair weather, a clear sky to the west, and a moderate southwest wind, the birds start, fly *west* along the Rhode Island coast going higher and higher in the air as they go west, and at or near Watch Hill go over the land and take a northwest course for the Great Lakes." This flight "begins about two hours before sunset, and on favorable occasions several flocks are in sight all the time, from twenty to two hundred in a flock. . . . Many leaders of flocks miss their bearings and turn up into Narragansett Bay. They then follow up to the head waters at the city of Providence, and follow a northwest course from there. In Col. J. H. Powel's List he says "this is the bird that comes from the east and flies to the west in its spring migration in May, from the 10th to the 20th, and is seen at no other time of year."

Mr. Dexter writes of the general Scoter migration, "There are seasons when circumstances of wind and weather are favorable in both spring and fall migration (in April and October), when the Scoters (*Americana deglandi* and *perspicillata*) pass very near our shores in vast numbers. On the 16th of last October, 1898,

I witnessed a flight past Sakonnet Point from 10 A. M. until 4 P. M. of an immense number. In the spring the same occurs, but not with the regularity of former years. This spring, 1899, on the 28th of April, I am told, a very large flight occurred, previous to which few birds had been seen."

The most interesting of the local bay migrations is that of the Cormorants (*Phalacrocorax carbo* and *dilophus*). During the fall and spring both the Common and Double-crested Cormorants migrate up and down the bay, chiefly by the Sakonnet River, from the "Cormorant Rocks" to the Kickamuit, Taunton and other rivers, where they feed. During the winter, after the Double-crested Cormorants have entirely or to a great extent left these waters, the Common Cormorant still follows this migratory movement, although to a less degree. Mr. Owen Durfee of Fall River writes us that this migration is affected by whether the herring are running in the Taunton River or not.

Land Birds.

The land-bird migration is as peculiar, and of as much interest, as that of the water birds. The spring migration along the coast seems to turn in somewhere on the Connecticut coast, cutting across through Providence, to the vicinity of Boston, and thence northward and records for arrival at these three places follow in order. The birds which breed in southern Rhode Island seem to work down as offshoots from this main migration, for arrival records for southern Rhode Island are invariably later than for Providence and vicinity. This is also known to be true of the migration at Fall River and Cape Cod region, which also seems to be cut off from the main migration route. There are also a number of local land migrations. The only one, however, worthy of note is that of the American Crow (*Corvus americanus*) which, like the Cormorant, during the winter months, feeds at low tide along the Kickamuit, Taunton and other rivers, and migrates at morning and evening through the Mount Hope lands, over Bristol promontory and Prudence Island, to a roost in the Greenwich woods.

The fall migration is so much more obscure and desultory, and there is such a lack of observations with regard to it, that anything definite cannot be stated, but from nocturnal observations made in Narragansett Bay and along the coast, it would seem that the main migration of land birds in the fall follows an outside ocean route.

A LIST OF THE BREEDING BIRDS OF RHODE ISLAND, 111 SPECIES.

Giving earliest dates on which nests contain eggs.

Pied-billed Grebe (*Podilymbus podiceps*)	June 6.
Common Tern (*Sterna hirundo*)	June 9.
Black Duck (*Anas obscura*)	May 5.
Blue-winged Teal (*Querquedula discors*)	?
Wood Duck (*Aix sponsa*)	May 10.
Ruddy Duck (*Erismatura jamaicensis*)	?
American Bittern (*Botaurus lentiginosus*)	May 23.
Least Bittern (*Ardetta exilis*)	May 23.
Green Heron (*Ardea virescens*)	May 21.
Black-crowned Night Heron (*Nycticorax n. nævius*)	May 7.
Virginia Rail (*Rallus virginianus*)	May 29.
Sora (*Porzana carolina*)	May 24.
Florida Gallinule (*Gallinula galeata*)	?
Woodcock (*Philohela minor*)	April 5.
Wilson's Snipe (*Gallinago delicata*) ?	[May 10.]
Spotted Sandpiper (*Actitis macularia*)	May 27.
Killdeer (*Ægialitis vocifera*)	May 10.
Piping Plover (*Ægialitis meloda*)	June 5.
Bob-white (*Colinus virginianus*)	May 25.
Ruffed Grouse (*Bonasa umbellus*)	May 8.
Mourning Dove (*Zenaidura macroura*)	May 15.
Marsh Hawk (*Circus hudsonius*)	May 25.
Sharp-shinned Hawk (*Accipiter velox*)	May 25.
Cooper's Hawk (*Accipiter cooperii*)	May 14.
Red-tailed Hawk (*Buteo borealis*)	April 5.
Red-shouldered Hawk (*Buteo lineatus*)	April 5.
Broad-winged Hawk (*Buteo latissimus*)	May 15.
American Sparrow Hawk (*Falco sparverius*)	May 10.

American Osprey (*Pandion h. carolinensis*) May 7.
American Long-eared Owl (*Asio wilsonianus*) April 10.
Short-eared Owl (*Asio accipitrinus*) ? (April 15).
Barred Owl (*Syrnium nebulosum*) March 15.
Screech Owl (*Megascops asio*) April 7.
Great Horned Owl (*Bubo virginianus*) February 28.
Yellow-billed Cuckoo (*Coccyzus americanus*) May 24.
Black-billed Cuckoo (*Coccyzus erythrophthalmus*) June 1.
Belted Kingfisher (*Ceryle alcyon*) May 15.
Hairy Woodpecker (*Dryobates villosus*) May 28.
Northern Downy Woodpecker (*Dryobates pubescens medianus*)
    May 12.
Red-headed Woodpecker (*Melanerpes erythrocephalus*) [July 28.]
Flicker (*Colaptes auratus*) April 29.
Whip-poor-will (*Antrostomus vociferus*) June 8.
Nighthawk (*Chordeiles virginianus*) June 5.
Chimney Swift (*Chætura pelagica*) June 10.
Ruby-throated Hummingbird (*Trochilus colubris*) May 20.
Kingbird (*Tyrannus tyrannus*) May 31.
Crested Flycatcher (*Myiarchus crinitus*) June 5.
Phœbe (*Sayornis phœbe*) May 7.
Wood Pewee (*Contopus virens*) June 15.
Least Flycatcher (*Empidonax minimus*) May 24.
Blue Jay (*Cyanocitta cristata*) May 2.
American Crow (*Corvus americanus*) April 13.
Bobolink (*Dolichonyx oryzivorus*) June 3.
Cowbird (*Molothrus ater*) May 16.
Red-winged Blackbird (*Agelaius phœniceus*) May 12.
Meadowlark (*Sturnella magna*) May 6.
Orchard Oriole (*Icterus spurius*) May 30.
Baltimore Oriole (*Icterus galbula*) May 24.
Purple Grackle (*Quiscalus quiscula*) April 25.
Bronzed Grackle (*Quiscalus quiscula æneus*) May 1.
Purple Finch (*Carpodacus purpureus*) May 19.
American Goldfinch (*Astragalinus tristis*) July 1.
Vesper Sparrow (*Poæcetes gramineus*) May 5.
Savanna Sparrow (*Ammodramus sandwichensis savanna*) May 17.
Grasshopper Sparrow (*Ammodramus savannarum passerinus*)
    June 5.

MIGRATION.

Sharp-tailed Sparrow (*Ammodramus caudacutus*)	May 24.
Seaside Sparrow (*Ammodramus maritimus*)	(May 31.)
Chipping Sparrow (*Spizella socialis*)	May 19.
Field Sparrow (*Spizella pusilla*)	May 19.
Song Sparrow (*Melospiza fasciata*)	May 10.
Swamp Sparrow (*Melospiza georgiana*)	May 24.
Towhee (*Pipilo erythrophthalmus*)	May 21.
Rose-breasted Grosbeak (*Zamelodia ludoviciana*)	May 23.
Indigo Bunting (*Cyanospiza cyanea*)	June 1.
Scarlet Tanager (*Piranga erythromelas*)	May 22.
Purple Martin (*Progne subis*)	May 25.
Cliff Swallow (*Petrochelidon lunifrons*)	May 31.
Barn Swallow (*Hirundo erythrogastra*)	May 18.
Tree Swallow (*Tachycineta bicolor*)	May 25.
Bank Swallow (*Clivicola riparia*)	May 26.
Cedar Waxwing (*Ampelis cedrorum*)	June 7.
Red-eyed Vireo (*Vireo olivaceus*)	May 28.
Warbling Vireo (*Vireo gilvus*)	May 30.
Yellow-throated Vireo (*Vireo flavifrons*)	May 24.
Blue-headed Vireo (*Vireo solitarius*)	May 25 to June 15.
White-eyed Vireo (*Vireo noveboracensis*)	June 2.
Black and White Warbler (*Mniotilta varia*)	May 21.
Blue-winged Warbler (*Helminthophila pinus*)	[May 30.]
Nashville Warbler (*Helminthophila ruficapilla*)	June 1.
Parula Warbler (*Compsothlypis americana usneæ*)	June 2.
Yellow Warbler (*Dendroica æstiva*)	May 25.
Chestnut-sided Warbler (*Dendroica pensylvanica*)	May 22.
Black-throated Green Warbler (*Dendroica virens*)	June 10.
Pine Warbler (*Dendroica vigorsii*)	May 25.
Prairie Warbler (*Dendroica discolor*)	May 27.
Oven-bird (*Seiurus aurocapillus*)	May 19.
Louisiana Water-Thrush (*Seiurus motacilla*)	May 15.
Maryland Yellow-throat (*Geothlypis trichas*)	May 2 to May 25.
Yellow-breasted Chat (*Icteria virens*)	May 31.
American Redstart (*Setophaga ruticilla*)	June 1.
Catbird (*Galeoscoptes carolinensis*)	May 25.
Brown Thrasher (*Harporhynchus rufus*)	May 19.
Carolina Wren (*Thryothorus ludovicianus*)	April 15.

House Wren (*Troglodytes aëdon*) May 27.
Short-billed Marsh Wren (*Cistothorus stellaris*) June 6.
Long-billed Marsh Wren (*Cistothorus palustris*) May 31.
White-breasted Nuthatch (*Sitta carolinensis*) April 20.
Chickadee (*Parus atricapillus*) May 9.
Wood Thrush (*Hylocichla mustelinus*) May 21.
Wilson's Thrush (*Hylocichla fuscescens*) May 26.
American Robin (*Merula migratoria*) April 25.
Bluebird (*Sialia sialis*) April 7.

NOTE: — ( ) indicate approximate date where data is lacking. [ ] indicate date of only nest taken. ? indicates lack of really good authentic data.

CORMORANT ROCK.
NEST OF COMMON TERN ON THE ROCK.

CORMORANT ROCK.

CORMORANT ROCK is situated one mile south from the most southeasterly point of the Island of Rhode Island. This jagged mass of weathered granite is about an acre in extent and rises some twenty-five feet above the level of the ocean. The highest point is nearest the northerly side, and the southerly exposure is broader and somewhat flattened. Separated from the larger rock by a deep, narrow channel, is another rock about one fifteenth as large, but nearly as high. Opening to the northwest, on the northerly side of the greater rock, is a small cove, filled with boulders, which is partly encircled by an arm of the rock that makes out to the north and west. It is only in this cove that a landing can be made with safety in smooth weather, for the ocean swell is constantly surging around the other sides in a foreboding manner. No land vegetation ever grows on this lonely rock, for whatever soil collects on the higher portions, through the disintegration of the rock or the deposits made upon it by birds, is swept away during heavy storms when the waves make a clean sweep of the rocks. The average rise and fall of the tide at this point is four feet, and the land around the rock, below the high tide line, is covered with a luxuriant growth of slimy ooze, rockweed (*Fucus*) muscles, and barnacles.

Looking at the rock from a geologist's point of view, it is of igneous origin, being composed throughout of gray granite of a coarse crystalline structure and well seamed with cleavage planes that divide it up in more or less regular parallelopipeds of varying size, from a few cubic inches to as many cubic yards. The rock rises abruptly in ten fathoms of water from a hard bottom, and forms a line with outcroppings of similar rocks that occur on Sakonnet Point and the southwesterly part of the Island of Rhode Island. The mesa top of the rock always presents a whitish appearance, which is owing to the lime deposited there by the birds that roost on its summit. This lime has penetrated the rock to a considerable distance in many places, having been carried there by the rain and sea water as they percolate through the joints.

Few places present a more desolate or foreboding appearance than this lonely rock after a cold spell of weather in the winter. At such times it is a mass of snow and granular salt water ice which freezes to a considerable thickness on all sides, where wave after wave throws its spray high in the air only to be blown on the rock to freeze and add one more layer to the rock's cold blanket. But it is at just such times as these that Cormorant Rock is a most fascinating place to the ornithologist, for then the shoal waters of Narragansett Bay become excessively cold, and parts that do not actually freeze over are full of drifting ice. Driven from the sheltered waters, the sea fowl make their way oceanward and seek some feeding ground where they can find a lee and a roosting place, for birds are fond of having some place of rendezvous that can be seen from a distance.

The most noticeable birds that are to be found on Cormorant Rock are those from which the rock is named. These large black birds may be seen from a distance, as one approaches, sitting majestically upon the highest parts, in groups of varying size, but never scattered over the entire rock. As one draws near, they take wing and fly off, often alighting in the water about a mile away. It is apparent that this rock has been the resort of these birds for a great many years. In an article by Mr. George H. Mackay, entitled "Habits of the Double-crested Cormorant (*Phalacrocorax dilophus*) in Rhode Island" and published in "The Auk," Vol. XI, No. 1, Jan., 1894, he says, — "These low lying black rocks have been in the past, and are still, the resort and roosting place of all the Cormorants living in and around these waters, and as they undoubtedly received their name many years ago from such occupancy it may be interesting to know that on a map dated July 20, 1776, which is in an atlas called the 'American Neptune,' published in London in 1776, and surveyed by Des Barres, that these identical rocks are correctly shown and located under the name of the 'Cormorant Rock.' It would not, therefore seem unreasonable to infer that they were so named on account of being frequented by these birds at that early period, or even before. If such a conclusion is admissible it would show an occupancy of certainly one hundred and sixteen years, and possibly for a longer period, as well known local names are preserved, when feasible, in order to avoid confusion. There is, however, other evidence of long occupancy of still greater interest to the ornithologist, in the fact that I discovered, on careful examination, that many of the projections of the rock on the mesa top, which afford good standing places, had apparently been worn smooth and glossy by long use." There are two species of Cormorants that frequent this rock, the Double-crested (*Phalacrocorax dilophus*) and the Common Cormorant (*Phalacrocorax carbo.*) The former being most common during the migrating season and the latter the permanent winter resident. Although a few of these birds may be seen around this rock at any time of the fall, winter and early spring, they can best be studied toward sundown when they come to it to roost for the night, from Narragansett Bay and its rivers. Before alighting, the first arrivals fly around the rock in a suspicious way and do not alight until they have encircled it several times, but those that come later, alight at once without hesitation, the presence of their companions who have already lit, no doubt inspiring them with confidence. During the winter months when the Cormorants frequent the rock, large numbers of gelatinous balls or pellets are to be found on the rock. They average about an inch in diameter and consist of a mass of vertebrae and other bones of fishes which the Cormorants are unable to digest and which they eject in this form. They are of particular interest as showing the food of the Cormorants. Mr. S. Garman of Harvard University has been good enough to examine a number of them and has identified the bones as belonging to Porgies (*Archosaugus probatocephalus* Walb.), and (*Micropogon undulatus* Linn.), young parrot fishes (*Labriods*), drums (*Scienoids*), and Crabs (*Cancer irroratus*). When the snow and ice have disappeared and the warm lengthening days of spring add that indescribable charm to all nature, the lonely Cormorants leave their winter home and follow the retreating ice line north, as far as Labrador. Here they breed and accustom their young to all the hardships of life in northern waters. But Cormorant Rock is not left long unoccupied. No sooner do the winter inhabitants leave, than a colony of Common Tern (*Sterna hirundo*), take up their abode and rear their young in the crevices of the hard rock. This colony consists of about 175 birds. They arrive in a large flock early in May and at once take possession of the rock. We have already observed that there is no land vegetation on the rock and as the nearest land is one mile distant the Terns do not attempt to bring material thence for their nests, but prefer to use the bleached bones of the fish that have been devoured by Cormorants and left to whiten in the crevices of the rock. These bones the Terns arrange (we can scarcely say weave) in small crevices of the rock, in such a way that they form a saucer-like depression and serve to prevent the eggs from rolling around on the rock. The Terns lay from one to five eggs which are hatched about the last of June; and the downy young may be found running nimbly over the jagged rock. When any one lands on the rock the Terns rise in a body with loud cries and circle around and around high over the rock. Occasionally one, more daring than the rest, darts downward toward the rock, uttering, as it does so, its sharp, piercing cry, and again, swooping upward, joins the excited throng. Terns will fly miles in search of the small fish on which they feed, and hence it is that one so often finds them on and about the fish traps and pounds in Narragansett Bay, where they may be seen sitting motionless side by side on the horizontal poles that are used to stiffen the vertical piles. It is reasonable to suppose that these birds come, for the most part, from the Cormorant Rock colony, for the only other headquarters in Rhode Island is Dyer's Island where some dozen pairs breed every year. (Auk, Vol. XIV, No. 2, p. 203.)

Among the occasional visitors to Cormorant Rock are the Turnstones (*Arenaria interpres*) which have been found there in the fall and spring, during their migrations. And it is surprising to find them so tame that one may approach within a few feet, before they take wing. Even then they rarely go far, never leaving the rock, but merely changing their position on its collar of seaweed and barnacles. They have never been seen to alight on the rock more than three or four feet from the water, and accordingly it would appear that they find food by the water's edge.

Still another visitor at the rock during the late spring and summer is the Spotted Sandpiper (*Actitis macularia*) which occasionally makes a "flying trip" off from Sachuest Point and awakens the stillness of this out of the way spot by its bright, clear whistle. It is not thought that these Sandpipers have any definite aim in coming to the rock, further than the fact that it makes an interesting spot to visit, when they feel vigorous, and like making a trip across one mile of intervening ocean. On a visit to the rock on the 18th of April, 1899, Mr. Sturtevant was suddenly surprised to see a Vesper Sparrow flit past and light on the rock only a few feet away. It appeared nervous and not at home, moving from place to place on the rock as if worried and unable to make up its mind to resume its migratory flight.

Among the most interesting of the smaller birds that make their way here are the Purple Sandpipers (*Tringa maritima*). When the winter winds are penetrating and bleak, these little fellows will stand motionless upon the cold rock just out of reach of the waves, and facing the wind, their backs arched and their heads drawn down upon their shoulders, they present a most cold and cheerless appearance. At such times they are remarkably tame, allowing one to approach within ten feet of them, without showing the least alarm. Finally, if one draws too near, they will fly off a few feet, or, more often, around to the other side of the rock, uttering as they do so their plaintive whistle.

We have devoted a chapter to this rock as it presents an ideal point of observation from which to study maritime avifauna. It is difficult to steal into the very environment of sea-birds as one can into that of land-birds, but hidden in one of the natural crevices of this rock, with the roar of the sea continually in one's ears, and with Stolid Sandpiper and shy Black Duck almost within reach, one feels an intruder, a traveller, as it were, in a foreign land.

A list of species observed upon the rock or immediately surrounding it is here given.

Holboell's Grebe (*Colymbus holbœllii*).
Horned Grebe (*Colymbus auritus*).
Loon (*Gavia imber*).
Red-throated Loon (*Gavia lumme*).

Great Black-backed Gull (*Larus marinus*).
American Herring Gull (*Larus a. smithsonianus*).
Bonaparte's Gull (*Larus philadelphia*).
Common Tern (*Sterna hirundo*).
Roseate Tern (*Sterna dougalli*).
Common Cormorant (*Phalacrocorax carbo*).
Double-crested Cormorant (*Phalacrocorax dilophus*).
Red-breasted Merganser (*Merganser serrator*).
Mallard (*Anas boschas*).
Black Duck (*Anas obscura*).
Old-Squaw (*Harelda hyemalis*).
American Eider (*Somateria dresseri*).
American Scoter (*Oidemia americana*).
White-winged Scoter (*Oidemia deglandi*).
Surf Scoter (*Oidemia perspicillata*).
Purple Sandpiper (*Tringa maritima*).
Spotted Sandpiper (*Actitis macularia*).
Turnstone (*Arenaria interpres*).
Vesper Sparrow (*Pooecetes gramineus*).

NOTE.

In the Annotated List in regard to the arrival and departure of each species — the date given in brackets — is approximate date when the species should arrive and depart in Rhode Island. The English name of the species following the Latin name, is the one authorized by the American Ornithologists' Union, the names following in italics are the local names of the species in Rhode Island, given in the order of their use and familiarity. Footnotes marked with an asterisk (*) refer to general records of species of no particular significance.

ABBREVIATIONS.

Auk. — The Auk, A Quarterly Journal of Ornithology, published for the American Ornithologists' Union, New York, L. S. Foster.

Bull. Nut. Orn. Club. — Bulletin of the Nuttall Ornithological Club, A Quarterly Journal of Ornithology, published by the Club, Cambridge, Mass.

F. & S. — Forest and Stream, A Weekly Journal of the Rod and Gun. Published by the Forest and Stream Publishing Company, New York.

Nid. — The Nidiologist, published by Harry Reed Taylor at Alameda, Cal., from Sept., 1893, to Feb., 1895, at New York from Mar. to Nov., 1895, The Nidologist at New York from Dec., 1895, to April, 1896, and at Alameda from May, 1896, to May, 1897.

O. & O. — Ornithologist and Oologist (formerly, The Oologist) published by Frank Blake Webster Company, Hyde Park, Mass.

Ran. Notes. — Random Notes on Natural History, published by Southwick and Jencks, Providence, R. I., No. 12 of Vol. II and Vol. III published by James M. Southwick, Successor.

Dr. Rives' List. — "The Birds of Newport," by William C. Rives, Jr. M. D. Proceedings of the Newport Natural History Society, 1883-4, page 28, Newport, R. I.

Col. Powel's List. — "List of Birds Shot near Newport," by Col.

John Hare Powel, Proceedings of the Newport Natural History Society, 1883-4, page 42, Newport, R. I.

Southwick's List. — "Our Birds of Rhode Island," by James M. Southwick, Proceeding of the Newport Natural History Society 1887-8, page 3, Newport, R. I.

Lawton's List. — "The Water-Birds of Newport, R. I.," by Charles H. Lawton, Proceedings of the Newport Natural History Society, 1887-8, page 16, Newport, R. I.

AN OSPREY'S NEST.
BRISTOL.
From " On the Birds' Highway."

PART II.

ANNOTATED LIST OF THE BIRDS OF RHODE ISLAND.

(1) 2. **Colymbus holbœllii (Reinh.).** HOLBŒLL'S GREBE. *Red-necked Grebe. American Red-necked Grebe.*— An irregular winter visitor. It is found in Narragansett Bay, though more commonly along the coast. Mr. Newton Dexter writes that in the Bay it is "common in October." Mr. G. W. Field writes that he "took two in the Seekonk River near Red Bridge in November, 1885." On account of its often being mistaken, by the local gunners, for the Red-throated Loon (*Gavia lumme*) and vice versa, notes in regard to it are somewhat untrustworthy. In Lawton's List[1] he says that "the American Red-necked Grebe .... is very rare." Two birds were seen, and one was shot, by Mr. Owen Durfee on the ice in the Taunton River on February 19, 1895.

(October) November to April.

(2) 3. **Colymbus auritus Linn.** HORNED GREBE. *Hell-diver, Little Diver, Tinker Loon, Tinker.*— A common winter resident along the ocean cliffs, rocky shores, beaches and in Narragansett Bay. It is without doubt our most common Grebe. In the late spring, for a week or so before going north, they seem to move off shore about a mile, and change of plumage at this time also takes place. Mr. J. M. Southwick tells me that Mr. Benjamin Earle, of Providence, took a Horned Grebe in full plumage in the latter part of May in Narragansett Bay.

October to April 24.

(3) 6. **Podilymbus podiceps (Linn.).** PIED-BILLED GREBE. *Dabchick.*— An uncommon summer resident, and a somewhat irregular visitant in the fall, found only to our knowledge, in the

---
[1] Lawton's List, p. 17.

brackish and fresh water ponds. It is often common in Easton's Pond, Newport, and in the ponds at Little Compton, and sometimes not uncommon more inland, having been taken at Peacedale, Cranston and other localities. Of late years the species seems to have become more uncommon, for now it certainly is not the most common of our Grebes as Lawton called it in his List.[1] Mr. Newton Dexter writes that it is "common in summer, breeding at Long Pond, Sakonnet Pt., and that he has seen old birds with young and has taken eggs." Mr. G. W. Field also writes that "a pair nested in Point Judith Pond in 1897."

May to October (April). Summer.

(4) 7. **Gavia imber (Gunn.).** LOON. *Big Loon.* — A common winter resident along the ocean beaches and cliffs, and is frequently seen in Narragansett Bay, where, however, it is less common than outside. Most of the northward migration is over in April, but they are seen, usually singly, in May still migrating. Crippled and barren birds remain not infrequently throughout the summer in our waters.

(September) November 16 to May 24 (June). July and August.

(5) 11. **Gavia lumme (Gunn).** RED-THROATED LOON. *Little Loon.* — An uncommon winter resident off the ocean cliffs and beaches, and occasionally seen in Narragansett Bay. During the summer small flocks of three or four birds, generally young, are often seen which, for want of better explanation, one may call crippled or barren birds. Mr. Newton Dexter writes that this species "passes the coast in fall and spring in great numbers." He took a bird in adult plumage in April, 1874, at Sakonnet Point, now in the Brown University collection, and he has since taken two others. Mr. J. M. Southwick also writes that there was a bird in full plumage taken at Prudence Island in the autumn of 1898 which is now in the Smith collection.

October 1 to April 27. June, July and August.

---

[1] Lawton's List, p. 17.

(6) 27. **Cepphus grylle** (Linn.). BLACK GUILLEMOT.— An extremely rare, irregular winter visitant. There are but three records for the species. A bird taken by Mr. Newton Dexter in January, 1859, in Mount Hope Bay. Mr. Dexter writes me that this "went to the Franklin Society at the time," but he believes it is not in existence now. One shot in Narragansett Bay, Bullocks Point, by Mr. S. W. Thayer of Pawtucket,[1] and a young male taken by Mr. Newton Dexter off Sakonnet Point, November 18th, 1892.

(7) 30. **Uria troile** (Linn.). MURRE.— There is a record of one bird being taken at Point Judith, and which is now in the possession of Mr. Silas Wright of Wakefield. Mr. F. T. Jencks is the authority for this record. Mr. Newton Dexter writes that "one was taken at Sakonnet in 1894."

(8) 31. **Uria lomvia** (Linn.). BRÜNNICH'S MURRE. *Sea Crow. Foolish Guillemot. Brunnich's Guillemot.* — An irregular winter visitant along the coast, and has been taken in Narragansett Bay,[2] near Bristol,[3] and off Warwick,[4] and in the Blackstone River, above Pawtucket, where, Mr. H. S. Hathaway informs me, a small flock were seen, and one was shot on December 16, 1894. This bird is now in the collection of Mr. Harry A. Cash of Pawtucket.

(9) 32. **Alca torda** Linn. RAZOR-BILLED AUK.— A rare winter visitant along the coast. It has been taken a number of times off Sakonnet Point, West Island, and Newport.[5]
November 18 to February 4.

(10) 34. **Alle alle** (Linn.). DOVEKIE. *Sea Dove.* — A rare irregular winter visitant, both along the coast and in Narragansett Bay. It has been taken at West Island, Sakonnet River, New-

---

[1] O. & O. Vol. 11, No. 1, p. 16.
[2] Ran. Notes, Vol. III, No. 2, p. 8.
[3] Dr. Rives' List, p. 41.
[4] Ran. Notes, Vol. I, No. 1, p. 6.
[5] Ran. Notes, Vol. III, No. 12, p. 91.

port, Riverside on the Providence River, and Mr. A. C. Bent of Taunton writes us that "one was taken at Rehoboth, Mass., on November 25, 1885," probably driven inland from the Rhode Island coast by severe weather. Mr. H. S. Hathaway, also writes that in November, 1893, "one was shot in Attleboro, Mass., in a mud puddle, formed by a wagon rut."

November to April 27.

(11) 36. **Stercorarius pomarinus (Temm.).** POMARINE JAEGER.— A bird was taken at Newport on October 9, 1892. The bird was mounted by Messrs. Southwick and Critchly of Providence, for Mr. J. M. K. Southwick of Newport. There is also a bird in the Smith collection, taken off Block Island, October 11, 1895. Mr. Newton Dexter writes that "they are rather common off shore in the summer."

(12) 40. **Rissa tridactyla (Linn.).** KITTIWAKE. — An uncommon winter visitant, most common in the fall. It being an off-shore species it is not uncommonly seen at Block Island, much less commonly off Narragansett Pier, and Charlestown Beach, and very rarely off Newport. Mr. Newton Dexter writes that they are very abundant off Point Judith in November at times, and he has noted them in large numbers off Sakonnet Point as early as September 15. Mr. F. T. Jencks writes that Messrs. Southwick and Jencks received several one winter taken at Wickford. Mr. Joseph Horton, who he believes got them, said they were common in the harbor.

(13) 47. **Larus marinus Linn.** GREAT BLACK-BACKED GULL.— A not uncommon winter resident along the coast, and occasionally seen in Narragansett Bay, though much less commonly.

August 26 to April 23.

(14) 51a. **Larus argentatus smithsonianus Coues.** AMERICAN HERRING GULL. *Sea Gull. Winter Gull.*— An abundant winter resident along the coast, and in Narragansett Bay and rivers. Barren and crippled birds remain all summer

in the Bay and can be seen on the Halfway Rocks off Prudence Island. The Gull 'Dick',[1] whose long life and history has interested so many observers, we are glad to be able to include among Rhode Island birds.

September 9 to May 16. June, July, and August.

(15) 58. **Larus atricilla** Linn. LAUGHING GULL. — Mr. Newton Dexter killed a bird at Sakonnet Point in September, 1884.

(16) 60. **Larus philadelphia** (Ord). BONAPARTE'S GULL. — A winter visitor, common in spring and fall along the coast, generally observed in small flocks. Out of a flock of ten, seven were shot near Warwick Neck Light by the keeper, May 18, 1888, and three young birds, Mr. H. S. Hathaway writes us, have been taken on the Seekonk River during three successive falls, the last on September 25, 1895. Mr. Newton Dexter also writes "that a bird was taken in the fall, years ago, in full plumage on the same river." Lt. Wirt Robinson says that he has "seen flocks in Newport Harbor of certainly two hundred individuals."

September 25 to May 18. There are also two August records for Newport, one on the 27th, and one for Westerly.

(17) 64. **Sterna caspia** Pallas. CASPIAN TERN. — An accidental visitant. Three records, a bird taken about September 1, 1878, at Brightman's Pond, near Noyes Beach by Mr. Wm. Gudgeon,[2] another, a female, taken at Westerly on July 27, 1881, by Mr. J. B. Dunn,[3] and still another reported to have been

---

[1] Auk, Vol. IX p. 227; Vol. X, p. 76; Vol. XI p. 73; Vol. XII p. 76; Vol. XIII p. 78; Vol. XV p. 49.
[2] Oologist Vol. 5, No. 4, p. 32.
[3] O. & O., Vol. 6, No. 6, p. 44.
Coues and Stearns's, New Eng. Bird Life, Part II, p. 357.

Note: (69) **Sterna forsteri** Nutt. FORSTER'S TERN. — In Coues and Stearns's New Eng. Bird Life "Mr. N. T. Lawrence speaks of two R. I. specimens (For. & Str., Vol. X, 1878, p. 235)." This is an error, for on looking up the record we find it reads "L. I." (Long Island) instead of "R. I." (Rhode Island).

taken on May 10, 1892, by Mr. Claude Dunn of Ocean View.[1] Mr. Newton Dexter writes, " have noted very many in the past ten years migrating from the north in August and September at Sakonnet Point. They usually fly high and are rarely killed."

(18) 70. **Sterna hirundo Linn.** COMMON TERN. *Mackerel Gull.*— A common summer resident, breeding on Cormorant Rock and Dyer's Island.[2] The nests on the Cormorant Rock are constructed of fish bones, as there is no vegetation. The bones are gathered together after the weather has washed them free from the Cormorant pellets with which the rock is strewn. (See Cormorant Rock.)

May 16 to September 20. One doubtful record for October 24, 1891. There is also a record of two being taken in October.[3]

(19) 72. **Sterna dougalli Montag.** ROSEATE TERN. *Mackerel Gull.*— An uncommon summer resident, not uncommon in August and September, straying probably from the Penikese Island colony, Mass., as it does not breed to our knowledge within the State.

(May) to (October).

(20) 74. **Sterna antillarum (Less.).** LEAST TERN.— One was taken at Bench Mark Rock, off Nayatt Point by Mr. Thomas Adcock in 1885. Others are said to have been taken. Mr. Newton Dexter writes that this species " was formerly common in Narragansett Bay, and reports the capture of one at Sakonnet on August 20, 1897, now in the Smith collection."

(21) 75. **Sterna fuliginosa Gmel.** SOOTY TERN.—An accidental visitant. Mr. Fred. T. Jencks took a specimen at Point Judith in September 1876.[4] Mr. H. S. Hathaway writes, that

---

[1] O. & O., Vol. 17, No. 6, p. 96.
[2] Auk, Vol. XIV, No. 2, p. 203.
[3] O. & O., Vol. 18, No. 10, p. 141.
[4] Osprey, Vol. II, Nos. 6 and 7, p. 91.
Birds of Connecticut by C. Hart Merriam, Conn. Acad. IV, 1887, p. 134.
Allen's Revised List Birds of Mass. Bull. 7. Am. Mus. of Nat. Hist., p. 227.
Coues and Stearns, New Eng. Bird Life, Part II, p. 374.

another, though he questions it, was found dead in Newport in 1877, and still another, a male, in full plumage, was taken at Woonsocket on July 16, 1897, now in the Smith collection.

(22) 77. **Hydrochelidon nigra surinamensis (Gmel.).** BLACK TERN. — An irregular spring and fall migrant. A fine adult was taken at Newport by Dr. Henry F. Marshall on August 16, 1880.[1] Mr. H. S. Hathaway writes that five, all young, and in fall plumage were shot at Sakonnet Point on August 24, 1891. Mr. Newton Dexter says that they are "very abundant off Sakonnet at times, usually after a heavy southeaster in early September."
August. September.

(23) 88. **Puffinus borealis Cory.** CORY'S SHEARWATER. *Grew Gull.* — An uncommon summer visitant. A bird was sent Dr. Wm. C. Rives, on September 30, 1886, by Mr. J. Glynn, Jr., of Newport,[2] and a bird was taken near Point Judith in October, 1886, which was stuffed by Mr. John Hague, and is now in collection of Brown University, and a male was taken at Newport on August 27, 1886, collection of Mr. R. L. Agassiz, now in the collection of the Museum of Comparative Zoology, Cambridge, Mass. The capture of these birds would tend to show, as Dr. Rives suggests, "that the flight of these birds extended as far west as the mouth of Narragansett Bay."[2] This was the year of that remarkable flight. Since then the following have been reported: Mr. A. C. Bent took a female near Cormorant Rock on October 26, 1890, one was taken in 1891 in Narragansett Bay, four were taken at Newport, October 9, 1892, and mounted for Mr. J. M. K. Southwick of Newport, by Messrs. Southwick and Critchly of Providence. Two were shot at Point Judith by Mr. N. N. Bishop and others, one on Aug. 15, 1894, which is in the collection of Mr. H. S. Hathaway, and the other a few days later, which is now in the Smith collection. Mr. Newton Dexter writes that "the bird has been taken off Block Island, and that he took one

---

[1] Bull. Nut. Orn. Club, Vol. V, No. 4, p. 237.
[2] Auk, Vol. V, No. 1, p. 108.

on September 28, 1898, off Warren's Point, West Island, and saw several others."

(24) 89. **Puffinus gravis (O'Reilly).** GREATER SHEARWATER. *Hagdon.* — Mr. Newton Dexter writes that he has "seen them off shore following mackerel fleets."

(25) 94. **Puffinus fuliginosus Strickland.** SOOTY SHEARWATER. *Black Hagdon.* — Mr. Newton Dexter obtained two specimens at Sakonnet Point in May, 1890, he writes, "one, a male, is now in Dr. C. T. Gardner's collection, the other had been picked by a frugal fisherman, but was fully identified." He has seen others at various times. During the great flight in August and September, 1886, which extended from Point Judith to Buzzard's Bay a few of this species were among the Cory's Shearwaters (*Puffinus borealis*). It is not unlikely therefore, that they were within Rhode Island waters at this time.[1]

(26) 106. **Oceanodroma leucorhoa (Vieill.).** LEACH'S PETREL. *Stormy Petrel. Mother Carey's Chickens.* — An uncommon transient visitor. Dr. Rives in his paper writes " are found .... the particular variety being probably Leach's Petrel." On June 9, 1889 a solitary bird was taken off Portsmouth Grove, and another bird was taken on October 14, 1891, in Narragansett Bay. On August 29, 1898, eight birds were seen together in Newport outer harbor. Mr. Newton Dexter calls the bird rather rare.

June 9 to October 14.

(27) 109. **Oceanodroma oceanicus (Kuhl.).** WILSON'S PETREL. — A not uncommon transient visitant. Mr. Fred T. Jencks took a male in spring plumage at Newport on August 2, 1880,[2] Mr. J. M. Southwick writes, Mr. G. M. Gray, of Providence, took ten or twelve off Narragansett Pier in July, 1881, and Mr. Newton Dexter writes that "it is very common off shore in July and August, often coming some distance up the Sakonnet River."

---

[1] Auk, Vol. IV, No. 1, page 71.
[2] Auk, Vol. V, No. 4, page 237.

Two were taken at Newport on August 4 and 8, respectively, 1899. Mr. Sturtevant and Mr. LeRoy King observed a great number, estimated in the thousands, off Point Judith, August 14, 1899.

(May) to (September).

(28) 115. **Sula bassana** (Linn.). GANNET. *Solan Goose.*
— A migrant off the coast, not uncommon at Block Island, but rarely seen near the coast line. Mr. H. S. Hathaway writes that two were taken on Narragansett Bay, October, 1891, and there is a specimen in the collection of the Newport Natural History Society, taken off Sachuest Point in 1891. Mr. Newton Dexter says that they " pass over the sea usually well off shore on their southern migration from the 10th of September, and later generally in flocks of from six to twenty, old and young together, and that they return in April, going north singly, sometimes two or three are in sight at once but never in close company. In October," he adds, "they are common between Brenton's Reef and Block Island."

October 26 to May 16, rare in winter.

(29) 119. **Phalacrocorax carbo** (Linn.). CORMORANT. *Common Cormorant. Shag.* " *Taunton Turkey.*" " *Taunton Shag.*" — A not uncommon winter resident in the vicinity of Cormorant Rock. Migrating up and down Narragansett Bay, generally by the Sakonnet River, to the Taunton, and other rivers to feed. This bird has been considered much rarer than it really is in Rhode Island. Cormorant Rock is probably about as far south as they winter in any numbers. There is a specimen taken in Narragansett Bay by the Rev. James Coyle (no other data) in the Newport Natural History Society collection. Mr. Geo. H. Mackay shot one bird at West Island on April 21, 1889, and one bird at Cormorant Rock on April 19, 1892. On October 26, 1890, Mr. A. C. Bent took three birds off Cormorant Rock, the one that is in his collection is *P. carbo*, and he believes without doubt the other two were the same. One bird was taken March 27, 1885, at Newport,[1] and one at Nayatt Point on

---

[1] Allen's Revised List Birds of Mass. Bull. Am. Mus. Nat. His., p. 229.

April 10, 1885,[1] Mr. Edward Sturtevant has taken them at Cormorant Rock, on the following dates, November 4, 1898, March 10 and April 24, 1899. Lt. Robinson writes that he observed this species frequently between 1888 and 1890 at Newport, and gives many dates of birds observed and taken between September 29 and June 3. Undoubtedly the small colony of about fifty to seventy-five birds that winter regularly about Cormorant Rock are *P. carbo*, no specimens of *P. dilophus* having ever been taken during the mid-winter months.

(See Cormorant Rock).

September 29 to June 3.

(30) 120. **Phalacrocorax dilophus (Swain.).** DOUBLE-CRESTED CORMORANT. *Shag.* "*Taunton Turkey.*" "*Taunton Shag.*" — A common fall and spring migrant. This species migrates like the preceding up and down the Sakonnet and Middle passages of Narragansett Bay to the Taunton and other rivers where they spend the days feeding. Mr. Newton Dexter writes that they sometimes migrate overland. (See Migration). A single bird, probably of this species rather than the preceding, was seen by Mr. Sturtevant flying east on August 9, 1899, over Brenton's Reef.

September 16 to November. April 22 to May 16.

(31) 129. **Merganser americanus (Cass.).** AMERICAN MERGANSER. *Goosander. Buff-breasted Merganser.* — An uncommon winter visitant, principally to the rivers and inland ponds. Col. J. H. Powell writes that he has taken only two near Newport, but that they are common in the Taunton River.

(November) to (March).

(32) 130. **Merganser serrator (Linn.).** RED-BREASTED MERGANSER. *Sheldrake. Common Sheldrake.* — A common winter resident along the coast, and in Narragansett Bay, often abundant outside on migrations. The birds are generally found near the rocky shores and cliffs, upon which they are often

---

[1] Ran. Notes, Vol. II, No. 3, page 23.

observed sitting. Crippled and barren birds remain throughout the summer.

October 29 to May 16, July and August.

(33) 131. **Lophodytes cucullatus** (Linn.). HOODED MERGANSER. *Hooded Sheldrake. Smew.* — A rare winter visitor, but not uncommon migrant. Mr. Newton Dexter writes "that a full plumaged male is very rarely seen."

(November) to (May).

(34) 132. **Anas boschas Linn.** MALLARD. *Wild Mallard. Green-head.* — An uncommon winter visitor along the coast and to Narragansett Bay and inland ponds. Mr. A. C. Bent writes "that he has seen specimens taken at Hundred Acre Cove, Barrington." There is a record of capture at Newport, November 1, 1875, where it is stated that they are very rare in the locality.[1] Mr. Southwick,[2] however, states in his List, (1887-8) that it "is not uncommon." A fine male was taken at Nayatt by Mr. R. H. Gibbs in spring of 1899.

October to (April).

(35) 133. **Anas obscura Gmel.** BLACK DUCK. *Dusky Duck.* — An abundant winter and rare summer resident, breeding sparingly. Both the *red-legged* and *green-legged* varieties are found, perhaps the *green-legged* the more commonly.

September 29 to April 24, a few in summer.

(36) 135. **Anas strepera Linn.** GADWALL. — The Widgeon (*Mareca americana*), and female Pintail (*Dafila acuta*) are so often taken for this species that it is difficult to secure any accurate data. There is a record for a female taken on February 26 at Newport by Mr. F. T. Jencks,[3] and there is a badly mutilated specimen in the Rhode Island College collection at Kingston, which was shot by Mr. John Hoxie at Carolina, May 4,

---

[1] F. & S., Vol. 5, No. 13, p. 204.
 Coues and Stearns, New Eng. Bird Life, Part II, p. 305.
[2] Southwick's List, p. 11.
[3] O. & O., Vol. 7, No. 15, p. 114.

1892. Lawton states in his list that it has been taken at Newport.[1]

(37) 137. **Mareca americana Gmel.** BALDPATE. *American Widgeon.* — A rare winter visitant. Mr. Sturtevant took a bird at Middletown on September 20, 1889. Mr. F. T. Jencks recorded the species " as unusually abundant in the waters of southern Rhode Island during November and the first week of December, 1882,"[2] and Mr. Newton Dexter writes that " they are not uncommon in the ponds along shore in the fall."

September 20 to (April).

(38) 139. **Nettion carolinensis Gmel.** GREEN-WINGED TEAL. — An uncommon migrant and rare winter visitant to the ponds, much the rarer of the Teals.*

October 7 to (April).

(39) 140. **Querquedula discors Linn.** BLUE-WINGED TEAL. — Not an uncommon migrant, rarer in the spring. Breeds locally. Mr. Newton Dexter writes that a nest with eggs which he saw was taken at Sakonnet in May, 1890, by a Mr. Sisson. Lawton says this species is " always to be met with after the first northwest wind in September."[3]

September, October, May.

(40) 142. **Spatula clypeata Linn.** SHOVELLER. *Shoveller Duck. Spoonbill.* — Lawton's List states that it has been taken at Newport.[1] Mr. Newton Dexter says that a pair were taken near Newport in 1858. One was killed at Quonocontaug Pond by Mr. E. W. Champlin, April 10, 1894. Mr. S. W. Field writes that the older natives say that it was one of the common Ducks that fed in Point Judith Pond.

---

[1] Lawton's List, p. 17.
[2] Bull. Nut. Orn. Club, Vol. VIII, No. 1, p. 62.
* F. & S., Vol. 15, No. 14, p. 271, and Vol. 17, No. 11, p. 211.
[3] Lawton's List, p. 16.

(41) 143. **Dafila acuta (Linn.).** PINTAIL. *Gray Duck. Sprig-tail.*— An uncommon migrant along the coast at Sakonnet, Newport, and Point Judith. Lawton's List states that this species "is occasionally met with round our bay, and is very abundant at Block Island."[1] Mr. F. T. Jencks, with Dr. H. F. Marshall, shot three at Point Judith on September 4, 1879.

September 4 to (April 15).

(42) 144. **Aix sponsa (Linn.).** WOOD DUCK. *Summer Duck.*— A regular migrant and uncommon summer resident. Mr. O. Durfee writes that "it used to breed and he believes still does about the Tiverton and Little Compton ponds." Lawton also speaks of its breeding in the "western part of the State."[1]

(March) to (November), uncommon in summer.

(43) 146. **Aythya americana (Eyt.).** REDHEAD. *Redheaded Duck.*— A rare migrant and winter visitant. Both Dr. Rives'[2] and Southwick's[3] List states that it has been taken, and Lawton[1] says that it "is very plentiful (?) in the salt pond on Block Island going out of the pond every night toward the west and returning at daybreak." Mr. Newton Dexter writes that a few are taken every fall at Sakonnet. One was killed at Pawtucket February 5, 1898.

(October) to (April).

(44) 147. **Aythya vallisneria (Wils.).** CANVAS-BACK. *Canvas-backed Duck.*— A very rare and irregular visitant. Dr. Rives,[2] Southwick[3] and Lawton[4] say that it has been taken at Newport. There is a record of four birds being taken at Point Judith, in November, 1881, by Mr. Fred. Skinner.[5]

October 2 to November.

---

[1] Lawton's List, p. 16.
[2] Dr. Rives' List, p. 40.
[3] Southwick's List, p. 11.
[4] Lawton's List, p. 17.
[5] F. & S., Vol. 15. No. 23, p. 417.

(45) 148. **Aythya marila (Linn.).** AMERICAN SCAUP DUCK. *Broad-bill. Blue-bill. Greater Scaup. Widgeon.* — Not an uncommon migrant and rare winter visitant. Mr. A. C. Bent writes "that they used to be found in large numbers off Bullock's Point, Providence River, in the late fall," and Mr. O. Durfee writes "that at Quicksand Pond, Little Compton, he should call the Blue-bills common migrants, if not winter residents." There is a record, stating that they were very plenty at Newport on October 27, 1880.[1] Lawton says "that they are quite abundant during the fall and winter migration."[2] and Mr. Newton Dexter writes that they are common in fall and spring in Narragansett Bay.

October to (May).

(46) 149. **Aythya affinis (Eyt.).** LESSER SCAUP DUCK. *Blue-billed Shoveller. Creek Broad-bill. Little Black-head Duck.* — Not an uncommon transient visitant. Lawton says[2] "that they are quite abundant" at Newport. There is a female in the collection of Brown & Nichol's School, Cambridge, Mass., taken on February 25, 1886, at Newport, by Mr. R. L. Agassiz, and one was taken at Field's Point, February, 1899, by Mr. E. H. Armstrong.[*]

(October 15 to November 25) February to (April).

(47) 150. **Aythya collaris (Donov.).** RING-NECKED DUCK. *Bastard Broad-bill.* — A very rare visitant. Col. J. H. Powel writes he has taken one, which was sent to the Academy of Natural Sciences, Philadelphia, Pa. Lawton states in his List[2] "that they are quite abundant in this vicinity (Newport) during the fall and winter migration," but we feel quite sure he must have mistaken the species.

(48) 151. **Clangula clangula americana (Bonap.).** AMERICAN GOLDEN-EYE. *Whistler. Whistler-wing. Golden-eye.* — A common winter resident in Narragansett Bay and tide

---

[1] F. & S., Vol. 15, No. 14, p. 271.
[2] Lawton's List, p. 16.
[*] O. & O., Vol. 6, No. 2, p. 14.

rivers, but uncommon along the coast except on migration, and when the Bay is frozen over.

(November 15) to April 20.

(49) 153. **Charitonetta albeola (Linn.).** BUFFLE-HEAD. *Butter Ball. Buffle-head Duck. Dipper.* — A not uncommon, and regular migrant along the coast and in Narragansett Bay. Mr. O. Durfee writes that it is "a regular migrant in the Little Compton Ponds."

October 13 to November 4. (April).

(50) 154. **Harelda hyemalis (Linn.).** OLD-SQUAW. *Long-tailed Duck. South-southerly.* — An abundant winter resident along the coast. Lawton calls it "our most common sea duck,"[1] which it is not, compared with White-winged or Surf Scoter.

November 7 to May 4.

(51) 155. **Histrionicus histrionicus (Linn.).** HARLEQUIN DUCK. *Harlequin.* — A rare winter visitant. Southwick includes it in his list as having been taken at Newport.[2] Lieut. Wirt Robinson writes that he saw "several on January 29, 1895, below Dutch Island toward Narragansett Pier." Mr. Newton Dexter took three at Sakonnet Point, and a young male in nearly adult plumage was shot at Narragansett Pier on December 28, 1893, now in Rhode Island College collection at Kingston.

December 29 to January 29.

(52) 160. **Somateria dresseri Sharpe.** AMERICAN EIDER. *Wamp. Eider Duck.* — A common winter resident and migrant along the coast. It is rarely seen in Narragansett Bay. Mr. Newton Dexter writes that "four specimens, a female and three young in first plumage were taken in Moswansicut Pond, Scituate."[*]

(53) 162. **Somateria spectabilis (Linn.).** KING EIDER.

---

[1] Lawton's List, p. 16.
[2] Southwick's List, p. 11.
[*] F. & S., Vol. XXIV, No. 12, p. 228.

— An irregular winter visitant. Dr. Rives says it "has been obtained this winter up the bay" (1884-5),[1] Lawton says[2] the bird is "now and then met with" at Newport. Mr. O. Durfee writes that he is informed that "this bird is tolerably common off Sakonnet Point in winter, one was shot there about the middle of February, 1899." Two were found in the collection of Mr. R. L. Agassiz, taken in December, 1885, at Newport now in the collection of the Museum of Comparative Zoology, Cambridge, Mass. Mr. H. S. Hathaway writes that a female was taken at Nayatt Point on November 27, 1879, one at Wickford in February, 1895, a male at Nayatt Point, January 1, 1884, probably the bird Dr. Rives referred to, and a male at Narragansett Pier about January 30, 1898, now in the collection.

November 27 to February.

(54) 163. **Oidemia americana Swains.** AMERICAN SCOTER. *Butter-bill. American Scoter Duck. Butter-bill Coot. Yellow-billed Coot.* — A common winter resident along the coast, not often met with in Narragansett Bay.

October to May 11.

(55) 165. **Oidemia deglandi Bonap.** WHITE-WINGED SCOTER. *Great May White-wing. May White-wing. Velvet Duck. Velvet Scoter.* — A common winter resident along the coast and in Narragansett Bay. The most common of the Scoters. Its peculiar May migration to the westward is spoken of at length under "Migration." Crippled and barren birds are seen throughout the summer, perhaps most commonly in the Bay.

September 15 to June 7. June, July and August.

(56) 166. **Oidemia perspicillata (Linn.).** SURF SCOTER. *Patch-poll Coot. Skunk-head. Surf Duck. Patch-bill Coot.* Females are called *Gray Coot.* — A common winter resident along the coast and in Narragansett Bay. Crippled and barren birds are seen during the summer months.

September 1 to May 14. July and August.

---

[1] Dr. Rives List, p. 40.
[2] Lawton's List, p. 16.

(57) 167. **Erismatura jamaicensis** (Gmel.). RUDDY DUCK. *Broad-bill. Booby.* — A not uncommon migrant along the coast and to the inland ponds, breeding locally. Lawton calls this bird the "most abundant of the migratory ducks, of which large numbers are shot at Easton's Pond every season."[1] In a note in "The Auk" by Mr. G. S. Miller, Jr., he speaks of a number of specimens in adult plumage, being taken during July at Sakonnet Point.[2] Mr. Newton Dexter writes that two broods were hatched in Long Pond, Sakonnet, in 1895; when about half grown the young disappeared.[*]

March and October. July.

(58) 169.1. **Chen cærulescens** (Linn.). BLUE GOOSE. A young bird was taken at Charlestown Beach on October 16, 1892, by Mr. F. L. Glezen and identified by Mr. Newton Dexter.[8] Mr. Dexter writes that he presented two, killed near Newport, to the Brown University collection.

(59) 169. **Chen hyperborea** (Pall.).? LESSER SNOW GOOSE. *Goose. Snow Goose. Mexican Goose.* — A rare and irregular winter visitant. Whether the birds recorded are of this race or *Chen hyperborea nivalis* cannot be determined.

(October) to (April).

(60) 172. **Branta canadensis** (Linn.). CANADA GOOSE. *Wild Goose.* — A common migrant. Three or four flocks comprising some two hundred birds were observed at Newport, bound south on December 17, 1898.[*]

March 13 to April 29. October.

(61) 173. **Branta bernicla** (Linn.). BRANT. *Brant Goose.* — A not uncommon migrant, and rare winter visitant. Lawton

---

[1] Lawton's List, p. 16.
[2] Auk, Vol. XVIII, No. 1, p. 118.
[*] F. & S. Vol. XVII, No. 11, p. 211.
[3] F. & S., Vol. XXX, No. 3, p. 48, and Providence Journal, Jan. 16, 1893. Prov. Journal, Jan. 25th, 1893.
[*] F. & S. Vol. XV, No. 20, p. 389.

states that it is "a regular spring migrant,"[1] and Dr. Rives[2] that it is "of extremely rare occurrence." Mr. J. M. Southwick writes that one was killed on the Austin farm at Exeter in April, 1894, and Mr. H. S. Hathaway, that a fine adult was shot at Sabin's Point, December 31, 1894. Mr. Newton Dexter writes that it is common off the coast in fall and spring migrations, and is frequently seen in Narragansett Bay.[*]

(November) to April 18.

(62) 180. **Olor columbianus (Ord.).** WHISTLING SWAN. *American Swan.* — Mr. Newton Dexter records in Forest and Stream[3] the capture of one bird by Mr. A. F. Stanton, about November 13, 1879, at Quonochontaug Pond, Westerly. Mr. Stanton is said to have seen two birds, but only one was taken. They were flying from the east. The skeleton of this bird was in the Brown University collection. Southwick's List speaks of a Trumpeter Swan being taken in the State, but he undoubtedly was referring to the above record.[4]

(63) 188. **Tantalus loculator Linn.** WOOD IBIS. — Mr. H. S. Hathaway writes that "a young bird was shot at Barrington on August 8, 1896, by Mr. Charles Miller. It was seen coming up the river in company with a large Heron, which, from description, must have been a Great-Blue (*Ardea herodias*). It was given to Dr. Nelson R. Hall by Mr. Miller and to Mr. William Mathewson by Dr. Hall, it has since been placed in the Smith collection."[5] There is also a record of one being taken at Seekonk, Mass., just over the Rhode Island border.[6]

---

[1] Lawton's List, p. 11.
[2] Dr. Rives List, p. 40.
[*] F. & S. Vol. XVIII, No. 6, p. 107.
[3] F. & S. Vol. XIII, No. 17, p. 848.
Allen's Revised List Birds of Mass. Bull. Am. Mus. Nat. Hist. Vol. 1, p. 233.
Coues and Stearns, New Eng. Bird Life, Part II, p. 297.
[4] Southwick's List, p. 4.
[5] Osprey, Vol. I, No. 5, p. 67.
[6] Auk, Vol. XIII, No. 3, p. 341.

(64) 190. **Botaurus lentiginosus (Montag.).** AMERICAN BITTERN. — A common migrant and uncommon summer resident. April 6 to November 10.

(65) 191. **Ardetta exilis (Gmel.).** LEAST BITTERN. — A common summer resident. It breeds in all the reedy marshes of the State.[1] There is a record for early arrival, March 1, 1881, at Providence,[2] and also a winter record for February 28, 1881.[3]
March 1 to September 3.

(66) 194. **Ardea herodias Linn.** GREAT BLUE HERON. *Crane.*— Common migrant, seen feeding during migrations along the rocky shores and about the inland ponds.
April 17 to 30. July 15 to October 16.

(67) 196. **Ardea egretta Gmel.** AMERICAN EGRET. — A bird was shot on Prudence Island, August 17, 1888,[4] in the collection of Brown University, another was shot by Mr. E. W. Champlin, in a cedar swamp, at Ocean View, on June 1, 1893,[5] and Mr. H. S. Hathaway writes that "after a severe northeast storm one was shot near Tiverton Four Corners on October 12, 1894, now the property of Mr. S. W. Williams of Providence," and one was also killed near Newport in 1896. Mr. Hathaway also writes "that an adult male has lately come into his possession, shot on the Sakonnet River at Tiverton, by an Italian farmer, on August 15, 1899. Mr. Newton Dexter writes in regard to a rumor of the possible breeding of this species in the State, " Five years ago this summer (1894) I noted a pair of Egrets in early June, which were every day to be seen about Narrow River and Wesquage Pond (Narragansett Pier). On the northwest side of the pond is a dense swamp, or was then, and the birds frequented that. The pond was then under the

---

[1] O. & O., Vol. 5, No. 10, p. 78.
[2] Nut. Orn. Club Bull., Vol. VI, No. 3, p. 186.
[3] O. & O., Vol. 6, No. 1, p. 8.
Coues and Stearns, New Eng. Bird Life, Part II, p. 274.
[4] O. & O., Vol. 14, No. 4, p. 63.
[5] O. & O., Vol. 18, No. 6, p. 94.

control of my friend, the late H. S. Bloodgood, who gave orders that the birds should not be disturbed or molested in any way. He informed me that in August he noted five birds there on several occasions, and was sure three of them were young birds, and that they were raised there. It certainly looks as if it were possible." On account of the uncertainty of this record, however, the species has not been included in the List of Breeding Birds.

(68) 200. **Ardea cærulea Linn.** LITTLE BLUE HERON.— A young bird, in white plumage, was taken at Warwick, on July 13, 1878.[1] Mr. H. A. Talbot records several Snowy Herons (*Ardea candidissima*) being seen on June 7, 1884, in the southern part of the State, but Mr. F. T. Jencks corrects this statement, and suggests that the birds he saw might have been Little Blue Herons.[2] Mr. Newton Dexter writes that a young male was taken at Sakonnet, August, 1892, now in Dr. Gardiner's collection, and he recalls an adult bird that was taken near Providence and mounted by Mr. John Hague, Taxidermist, about twenty-five years ago. Mr. H. S. Hathaway writes that Mr. J. W. Critchley, Taxidermist, "had a young bird sent in to him on July 24, 1899, from Wakefield by Mr. Silas Wright."

(69) 201. **Ardea virescens Linn.** GREEN HERON. "*Fly-up-the-creek.*" — A common summer resident.
April 20 to September 10 (October).

(70) 202. **Nycticorax nyticorax nævius (Bodd.).** BLACK-CROWNED NIGHT HERON. *Night Heron. Shitepoke. Quwark.*— A common summer resident, wintering rarely, certainly very much less commonly than in Massachusetts. Formerly there was a large herony near Mount Hope, and Mr. O. Durfee writes that "as late as 1894 there was a large one on Prudence Island,"

---

[1] Bull. Nut. Orn. Club, Vol. V, No. 2, p. 123.
Allen's Revised List Birds of Mass. Bull. Am. Mus. Nat. Hist., p. 235.
[2] O. & O., Vol. 9, No. 7, p. 80 and No. 8, p. 103.

which Mr. F. W. Field says still exists, but much depleted. The inhabitants slaughtering the birds to "feed the pigs."

April 1 to September 23, also in winter, January.[1]

(71) 203. **Nycticorax violaceus** (Linn.). YELLOW-CROWNED NIGHT HERON. — A male was shot by Mr. Charles H. Kennedy on April 23, 1886, in Tiverton, less than a third of a mile south of the Fall River line.[2] A young female was taken in August, 1892, at Newport, by Mr. J. Livermore.[3]

(72) 205. **Grus canadensis** (Linn.). LITTLE BROWN CRANE.—A straggler from the West. Mr. Benjamin Burlingame took one October 9, 1889, at Natick Hill, now in possession of Mr. J. M. Nye,[4] of River Point. Two were reported seen.

(73) 208. **Rallus elegans** Aud. KING RAIL. — There are but three records for this species. One killed at Wakefield on February 12, 1889, by Mr. N. R. Potter. This bird was mounted for Mr. Frank Phillips by Mr. J. M. Southwick. It was shot near a partly open brook, the snow along the brook being about five inches deep and covered with its tracks, and Mr. Potter says he killed one near the same place in the fall of 1888. A fine male was taken at Newport, January 21, 1896, and is now in Mr. H. S. Hathaway's collection.

(74) 212. **Rallus virginianus** Linn. VIRGINIA RAIL. *Red-breasted Rail.?* — A common summer resident and migrant. It has been observed as late as November 2.[5]

(April) to November 2.

(75) 214. **Porzana carolina** (Linn.). SORA. *Carolina Rail.*

---

[1] Ran. Notes, Vol. I, No. 1, p. 9, and No. 2, p. 8.
[2] Ran. Notes, Vol. III, No. 7, p. 49.
[3] Auk, Vol. XI, No. 2, p. 177.
[4] Auk, Vol. VII, No. 1, p. 89. O. & O., Vol. 14, No. 10, p. 15. F. & S. Vol. XXXIII, No. 19, p. 264.
Independent Citizen, Providence, Nov. 16, 1889.
[5] Dr. Rives' List, p. 36.

— A common migrant and summer resident, the most common of our Rails.

(April) to October 12.

(76) 215. **Porzana noveboracensis (Gmel.).** YELLOW RAIL. — There are four records for this species. Mr. F. L. Glezen, of Providence, shot a bird at Charlestown Beach on September 28, 1886,[1] and a specimen was received from Mr. C. H. Lawton by Mr. J. M. Southwick on September 23, 1887, shot by a Newport gunner. One was killed by flying against a telegraph wire in Cranston. One was taken September 14, 1894, (exact locality unknown) by Mr. H. O. Havemeyer, Jr.

(77) [217]. **Crex crex (Linn.).** CORN CRAKE. — A straggler from Europe. Mr. Newton Dexter killed one in Cranston in 1857, and it is now in the Franklin Society collection at Providence.[2]

(78) 218. **Ionornis martinica (Linn.).** PURPLE GALLINULE. — An accidental visitant. A bird was taken at Westerly, in 1857?,[3] and another in 1875 by Mr. Newton Dexter. One at Warwick about August, 1886, and lived at least a year in Handy's Dime Museum, Providence,[4] one was found dead on January 13, 1889,[5] now in Dr. Gardiner's collection. Mr. Newton Dexter writes that it was picked up dead on Mr. Warren Kempton's farm, and was mounted by him in crude fashion. Mr. Dexter obtained it of him and remounted it. Another was killed at Sakonnet in the fall of the same year by Mr. Newton Dexter, another flew aboard a schooner at Wilkesbierre Pier, Providence, May 13, 1890, and was brought to Mr. J. M. Southwick for identification by Mr. G. F. Snow. A male was shot at

---

[1] O. & O., Vol. 12, No. 2, p. 32.
[2] Ran. Notes, Vol. I, No. VI, p. 3.
Allen's Revised List Birds of Mass." Bull. Am. Mus. Nat. Hist., p. 265.
[3] Bull. Nut. Orn. Club, Vol. VII, No. 2, p. 124.
Allen's Revised List Birds of Mass. Bull. Am. Mus. Nat. Hist., p. 236.
Coues and Stearns, New Eng. Bird Life, Part II, p. 293.
[4] Ran. Notes, Vol. III, No. X, p. 79.
[5] F. & S., Vol. XXXIII, No. 19, p. 364.

Newport about May 24, 1893, and mounted for the Newport Natural History Society collection.

(79) 219. **Gallinula galeata** (Licht.). FLORIDA GALLINULE. — A not uncommon migrant and local summer resident. Mr. Newton Dexter says "it is common at Long Pond, Sakonnet, in summer and fall. Breeds there, have taken eggs and young birds." Lt. Wirt Robinson in October, 1888,[1] took two, and saw at least a dozen others, and saw one at Almy's Pond on October 10, 1889.*

(May). September 10 to November.

(80) 221. **Fulica americana** Gmel. AMERICAN COOT.— *Coot. Marsh Hen. Mud Hen.* — A not uncommon migrant, to the fresh water ponds, sometimes fairly abundant. Mr. Newton Dexter writes that it is very abundant in latter part of September and through October at Sakonnet. Mr. Howe took a male, September 6, 1898, off Jamestown, Newport outer harbor, and Mr. H. S. Hathaway writes that he has a bird in his collection, taken at Field's Point, October 15, 1894.

(April) September, October 15.

(81) 222. **Crymophilus fulicarius** (Linn.). RED PHALAROPE. — An uncommon migrant. Lt. Wirt Robinson took one at Newport, on October 11, 1888;[1] another, in full plumage, was taken at Sakonnet on August 26, 1889; and another October 26, 1887, at the same place, and one at Newport, September 27, 1890.[2] Other records are one taken at Ocean View, May 24, 1892, one at Newport on May 23, of the same year, and one at Gaspee Point in the fall of 1898.

May, October.

(82) 223. **Phalaropus lobatus** (Linn.). NORTHERN PHALAROPE. — A not uncommon migrant. Dr. Rives records the cap-

---

[1] Auk, Vol. VI, No. 2, p. 194.
* F. & S., Vol, XV, No. 19, p. 371 and Vol. XVII, No. 11, p. 211.
[2] O. & O., Vol. 15, No. 11, p. 116.

ture of a bird at Newport on August 30, 1876, by Mr. F. W. Rhinelander.[1] Mr. F. T. Jencks took one at Point Judith, September 5, 1879, Mr. Sturtevant took two on Sakonnet River, May 15, 1892, and saw five on September 4, 1899, near Gull Rock, Narragansett Bay. There is one taken in August, 1894, and a female on May 11, 1898, at Newport, now in the Smith collection. Mr. Newton Dexter writes "a remarkable flight of these birds occurred at Sakonnet on May 15, 1895. A heavy southeast gale was blowing with much rain. Hundreds of flocks passed over the land at a point near the breakwater flying southeast. From a dozen to fifty in a flock."

May, August 16 to September 15.

(83) 224. **Steganopus tricolor Vieill.** WILSON'S PHALAROPE. — Dr. Rives records the capture of a bird taken by Mr. F. T. Jencks at Newport on August 2, 1880.[1] There is also a female taken at Newport on August 20, 1883, in the collection of the Boston Society of Natural History. Mr. J. Glynn, Jr., secured an immature bird from a local sportsman of Newport, on September 13, 1886.[2] Mr. Newton Dexter took one at Sakonnet, August 24, 1899, now in Dr. G. T. Gardiner's collection, which also contains another specimen.

(84) 228. **Philohela minor (Gmel.).** AMERICAN WOODCOCK. — A common migrant, and formerly common summer resident, now becoming almost rare as a breeding bird. There is a record for very early nesting, April 16, young having just left the nest.[4] [*]

(March) to December 4.

(85) 230. **Gallinago delicata (Ord.).** WILSON'S SNIPE.

---

[1] Dr. Rives' List, p. 39.
[2] Dr. Rives' List, p. 39.
Bull. Nut. Orn. Club, Vol. V, No. 4 p. 237.
Coues and Stearns, New Eng. Bird Life, Part II, p. 187.
[3] Auk, Vol. IV, No. 1, p. 73.
[4] Ran. Notes, Vol. II, No. V, p. 8.
[*] F. & S., Vol. XV, No. 19, p. 371, Vol. XXI, No. 25, p. 498. Vol. XXXV, No. 16, p. 312.

English Snipe. Common Snipe. — A common migrant to the fresh, and not uncommonly to the salt water marshes. This species has been taken in December and January. Mr. W. Hare H. Powel writes that "at first the birds are found where fresh and salt water meet, and as the season advances they work back to the more upland ponds, springs and slews, especially so in the autumn."

March 9 to April 30.   September 15 to November 1.   December and January.

(86) 231. **Macrorhamphus griseus (Gmel.).** Dowitcher. *Red-breasted Snipe. Brownback. "Deutscher" German Snipe. Fool Plover.* — A not uncommon migrant.*

(May).   July 5 to October 8.

(87) 232. **Macrorhamphus scolopaceus (Say).** Long-billed Dowitcher. — Mr. Sturtevant took a female at Middletown on October 8, 1890 now in the collection of Mr. William Brewster, Cambridge.

(88) 233. **Micropalama himantopus (Bonap.).** Stilt Sandpiper. *Mongrel. Bastard Yellow-leg.* — An uncommon spring, but not uncommon fall migrant. Mr. F. T. Jencks took a male in the spring plumage on August 2, 1880, at Newport,[1] and another at Point Judith on September 5, 1879. Mr. H. S. Hathaway writes "that Mr. Newton Dexter took one at Sakonnet Point on May 9, 1895, now in the Smith collection." Mr. LeRoy King took a female on August 1, 1899 and two on August 13, 1899 at Middletown. Mr. F. T. Jencks writes that they are not uncommon at Point Judith in the fall. Mr. Newton Dexter writes "that he has taken a dozen this season." (August 1899).

May, August 1 to September 5.

(89) 234. **Tringa canutus Linn.** Knot. *Robin Snipe.*

---

* F. & S., Vol. VI, No. 23, p. 376.
  Bull. Nut. Orn. Club. Vol. V, No. 4, p. 237.

Red-breasted Sandpiper. — A common migrant. Dr. Rives calls it "a somewhat rare species" at Newport.[1]

(May 15) to (June 10), July 11 to (November).

(90) 235. **Tringa maritima Brünn.** PURPLE SANDPIPER. — A common winter resident on Cormorant Rock, and during severe weather found on Sachuest Point. Mr. H. S. Hathaway writes that "Dr. H. F. Marshall shot four on the rocks off Newport, November 27, 1879."

September 13 to February 5 (March). One late record, May 15.

(91) 239. **Tringa maculata Vieill.** PECTORAL SANDPIPER. *Kreiker. Creaker. Grass Snipe. Pert.* — An abundant migrant. Found on the small salt marshes on Narragansett Bay, as well as on those along the coast. Mr. H. S. Hathaway writes that one, a female, was shot at Hammond's Pond, Pawtucket, on September 13, 1894.

(Spring, rare) July 16 to October 14 (November).

(92) 240. **Tringa fuscicollis Vieill.** WHITE-RUMPED SANDPIPER. — An uncommon fall migrant associating with *Tringa minutilla* and *Ereunetes pusillus.* Mr. Howe has taken it at Jamestown and Middletown.

July 11 to (October).

(93) 241. **Tringa bairdii (Coues).** BAIRD'S SANDPIPER. — Mr. H. S. Hathaway writes that "one was shot at Point Judith, September 4, 1892, by Mr. I. B. Mason's son, who has it in his possession. Mr. J. M. Southwick writes that there is one in the Smith collection, taken August 26, 1895, at Sakonnet.

(94) 242. **Tringa minutilla Vieill.** LEAST SANDPIPER. *Peep. Wilson's Sandpiper.* — An abundant migrant to the salt marshes, beaches and rocky shores along the coast, bays and tide rivers.

(April 25) to May 18, July 15 to October.

---

[1] Dr. Rives' List, p. 36.

(95) 243a. **Tringa alpina pacifica (Coues).** RED-BACKED SANDPIPER. *Winter Snipe.* — A not uncommon fall migrant. A bird was taken October 3, 1890, on the Second Beach marshes, Middletown, and Mr. A. C. Bent writes that they were " common at Hundred Acre Cove, Barrington on August 29, 1891, noted fifty there." They have been taken at Sakonnet and Point Judith.

(April) to (May 30) August 29 to October 20.

(96) 246. **Ereunetes pusillus (Linn).** SEMIPALMATED SANDPIPER. *Peep.* — An abundant migrant, associating with *Tringa minutilla.*

May 13 to June 2, July 15 to October 10.

(97) 247. **Ereunetes occidentalis Lawr.** WESTERN SEMIPALMATED SANDPIPER. — An uncommon migrant; as it occurs with both species of " peep," *T. minutilla* and *E. pusillus*, it is often overlooked. Mr. LeRoy King took one on August 25, 1899, and Mr. Sturtevant took one on August 29, 1899, at Middletown.

(May) (July to October).

(98) 248. **Calidris arenaria (Linn.).** SANDERLING. *Sanderling Sandpiper.* — A common migrant along the ocean beaches. Mr. H. S. Hathaway writes that "one was shot on Hammond's Pond, Pawtucket, between September 6 and 9, 1892," this is the only inland record.

March 20 to June. August 1 to September 27.

(99) 249. **Limosa fedoa (Linn.).** MARBLED GODWIT. *Common Marlin.* — A rare migrant. Dr. Rives states that they are "met with rarely" at Newport.[1] Dr. H. F. Marshall killed two, one at Newport, and one at Westerly. Mr. Newton Dexter " writes that they are very rare."

(August, September).

(100) 251. **Limosa hæmastica (Linn.).** HUDSONIAN GODWIT. *Ring-tailed Marlin.* — A rare migrant, a few, however, are

---

[1] Dr. Rives' List, p. 37.

killed each year. Dr. Rives states as of the preceding species that it is "met with rarely" at Newport.[1]

August, September.

(101) 254. **Totanus melanoleucus (Gmel.).** GREATER YELLOW-LEGS. *Winter Yellow-leg. Great Yellow-leg.* — A common migrant to the ocean marshes, but rarely to the marshes of Narragansett Bay, though frequently seen at inland ponds. Mr. O. Durfee writes that one was "reported on good authority on March 10, 1899, at Westport Harbor," just over the Rhode Island line.

April 10 to May 24, August 8 to October 15.

(102) 255. **Totanus flavipes (Gmel.).** YELLOW-LEG. *Lesser Yellow-leg. Summer Yellow-leg.* — A common fall migrant to the ocean and Narragansett Bay marshes. One spring record.

April 28, July 3 to September 28.*

(103) 256. **Helodromas solitarius (Wils.).** SOLITARY SANDPIPER. — An uncommon migrant. It has been taken at Easton's Pond, Newport, and on the Second Beach marshes, Middletown, and at Sakonnet, as well as inland. Mr. F. T. Jencks writes that he has "observed this species during the summer several times at Mashapaug Pond, Providence."

May 3 to 18, July 18 to September 25.

(104) 258. **Symphemia semipalmata (Gmel.).** WILLET. — A rare spring, and uncommon fall wandering visitant. Dr. Rives states that at Newport it is "a comparatively rare species."[1] Mr. O. Durfee writes that "on September 2, 1882, he took one, and saw another at Quicksand Pond, Little Compton." Mr. H. S. Hathaway reports the capture of one at Newport in the fall of 1897, by Mr. H. Havemeyer. Mr. LeRoy King of Newport killed a female, and saw a market hunter with two others, on August 12, 1898, taken on the Second Beach marshes,

---

[1] Dr. Rives' List, p. 37.
* F. & S. Vol. 6, No. 23, p. 376.

Middletown. These birds were the only ones heard of or seen that season. Mr. G. W. Field says that several are shot annually on the Point Judith marshes.

(May) August 12 to September 2.

(105) 261. **Bartramia longicauda (Bechst.).** BARTRAMIAN SANDPIPER. *Grass Plover. Upland Plover.* — Not an uncommon migrant, probably breeds. This species was formerly much more common in the State than it is now, Prudence Island being a famous shooting ground for them. Dr. Rives states "the much sought after Grass Plover, — now no longer common here (Newport)" etc., showing that early in the '80's it had grown rare.[2] On the uplands of Conanicut Islands, near Beavertail Light they are perhaps the most common of anywhere along the coast.

(April), August 14 to (September).

(106) 262. **Tryngites subruficollis (Vieill.).** BUFF-BREASTED SANDPIPER. — Mr. Newton Dexter states "that he has taken one or two nearly every season." There is one he took in September, 1896, now in the Smith collection.

(107) 263. **Actitis macularia (Linn.).** SPOTTED SANDPIPER. *Tip-up. Peet-weet.* — An abundant summer resident throughout the State.

April 18 to September 13. (October 20).

(108) 264. **Numenius longirostris Wils.** LONG-BILLED CURLEW. *Sickle-bill.* — Mr. Newton Dexter writes that "the last one he killed or has seen in Rhode Island was about 1862." A bird was taken by Mr. Thomas R. Stetson, at Round Swamp, Jamestown, on September 9, 1897, which was mounted by Mr. J. W. Critchley, Taxidermist, and now is in the collection of Mr.

---

[1] Dr. Rives' List, p. 36.
[2] Dr. Rives' List, p. 37.

William Brewster of Cambridge.[1] Dr. Rives states that at Newport they have this species.[2]

(109) 265. **Numenius hudsonicus** Lath. HUDSONIAN CURLEW. *Jack Curlew.* — An uncommon migrant to coast marshes. Mr. F. T. Jencks writes that "he shot one at the mouth of the Pawtuxet River about 1876."

(May) July to September 12.

(110) 266. **Numenius borealis (Forst.).** ESKIMO CURLEW. *Esquimaux Curlew. Doughbird.* — A very rare migrant in the fall. Formerly quite abundant migrating with the *Charadrius dominicus*. Dr. Rives[2] states that if he is "not mistaken," it has been taken at Newport. Mr. H. S. Hathaway writes that "Dr. H. F. Marshall killed six at Little Compton in 1886.

(September).

(111) 270. **Squatarola squatarola (Linn.).** BLACK-BELLIED PLOVER. *Beetle-head. Black-breast. Frost-bird.* — A not uncommon migrant. There has been a noticeable increase in their number during the last two seasons.

(May) August 9 to September 12 (October 15).

(112) 272. **Charadrius dominicus** Müll. GOLDEN PLOVER. *Green-head. Muddy-breast.* — An uncommon migrant, apparently becoming scarcer and scarcer each season. Mr. Newton Dexter writes "that they were very abundant formerly in August."

(May), August 28 to October 15 (November 10).

(113) 273. **Ægialitis vocifera (Linn.).** KILLDEER. *Killdeer Plover.* — An uncommon migrant and rare summer resident. This species has been reported from almost all parts of the State, Providence, Newport, Middletown, Wickford, Drownville, Kingston, Bristol, Warwick, Cranston, Conanicut Island, Point Judith, Sakonnet, Little Compton, Watch Hill, Block Island, and Coles

---

[1] Auk, Vol. XVI, No. 2, p. 189.
[2] Dr. Rives' List, p. 37.

River. There was a great flight of them in November, 1888, along the New England coast.[1] "Mr. Charles Doe took a set of four eggs at Wickford, May 11, 1894, and another at Cranston, May 30, 1896. Mr. H. A. Talbot reported them breeding at Warwick,[2] and Mr. F. E. Newbury, of Providence, found a pair nesting at Warwick in 1894 and 1895.[3] Mr. F. T. Jencks writes that a pair bred in Drownville in 1899.*

March, April, May, August, September, November, December, January and February.

(114) 274. Ægialitis semipalmata Bonap. SEMIPALMATED PLOVER. *Ring-neck. Little-ring Plover. Ring Plover.*— An abundant migrant to the coast and Bay beaches and marshes.

April 19 to May 15, July 6 to September 25.

(115) 277. Ægialitis meloda (Ord.). PIPING PLOVER. *Ring-neck.* — Not an uncommon migrant, and uncommon summer resident. Mr. O. Durfee writes that "a few breed along the Westport and Little Compton beaches," and Mr. H. S. Hathaway writes that "an adult and two young were shot at Charlestown Beach in June, 1895," now in the Smith collection.

(March) June, August (September).

(116) 283. Arenaria interpres (Linn.). TURNSTONE. *Rock Plover. Brant-bird. Horse-foot Snipe. Calico-bird.* — A common migrant to the rocky shores, and Cormorant Rock.

May, August 4 to September 13. There is a questionable record of four birds reported from Middletown on January 7, 1891.

(117) 289. Colinus virginianus (Linn.). BOB-WHITE. *Quail. American Quail.* — A common, and in some seasons, abundant resident. Reservations in different parts of the State have from time to time been established and stocked. There is

---

[1] Auk, Vol. VI, No. 3, p. 255.
[2] O. & O., Vol. 9, No. 5, p. 58.
[3] Nid., Vol. 8?, No. 3, 4, 5, p. 43.
* F. & S., Vol. XXIV, No. 12, p. 249, and Vol. XXVIII, No. 12, p. 225.

a record of a nest with ten eggs being found on October 10, 1894, near Wakefield by Mr. E. O. Schuyler.[1]

(118) 300. **Bonasa umbellus (Linn.).** RUFFED GROUSE. *Partridge.* A common resident in the northern and western portions of the State. Absent as far as our knowledge extends on Bristol promontory, and all the islands of Narragansett Bay.

(119) 316. **Zenaidura macroura (Linn.).** MOURNING DOVE. *Long-tailed Dove. Carolina Dove* — A not uncommon summer resident in the northern and western portions of the State. At Bristol, and on all the islands in Narragansett Bay, except perhaps on the Island of Rhode Island it is rare or absent.*
April 3 to October 6.

(120) 325. **Cathartes aura (Linn.).** TURKEY VULTURE. — A rare straggler from the South. Mr. H. S. Hathaway writes that a bird was taken at Niantic in the summer of 1861, which was mounted and presented to the Franklin Society collection by Mr. Newton Dexter. Another was taken in November, 1890, at the northeast end of Conanicut Island by Mr. E. D. Arnold, son of the lighthouse keeper. Mr. LeRoy King of Newport secured a specimen that was shot by a workman on June 20, 1893, on the King Farm, Brenton's Point, Newport. At the time it was feeding on some kind of a dead animal, probably a cat. The bird was mounted by Messrs. Southwick and Critchley and is now in Mr. King's possession. One, which is now in the Smith collection, was brought to Mr. J. W. Critchley, Taxidermist, about May 10, 1896 (exact locality unknown).

(121) 331. **Circus hudsonius (Linn.).** MARSH HAWK. — A common migrant and summer resident. This species is fast becoming rare.
(March 15) April 17 to October 29.

---

[1] Prov. Journal, Oct. 27, 1894.
* Col. Powel's List, p. 42.

(122) 332. **Accipiter velox (Wils.).** SHARPED-SHINNED HAWK. — A rather uncommon migrant, and summer resident. Probably occurs rarely in winter.

(April) to (November).

(123) 333. **Accipiter cooperii (Bonap.).** COOPER'S HAWK. *Chicken Hawk.* — A common migrant and summer resident, probably occurring rarely in winter.

(April) to (November).

(124) 334. **Accipiter atricapillus (Wils.).** AMERICAN GOSHAWK. — An irregular and rare winter visitant. Mr. F. T. Jencks writes that "large numbers were killed in a winter about twenty-five or twenty years ago." Mr. H. S. Hathaway writes that "Mr. Patrick Wally shot an adult female at Scituate, on October 30, 1893, and that during the winter of 1896 twenty-nine birds, twenty-two adults, and seven young were brought in from nearby towns in Rhode Island and Connecticut to Mr. J. W. Critchley, Taxidermist, to be mounted, one of which he bought, a female shot at Scituate, on November 24, 1896."[1] Mr. O. Durfee also writes that he saw "one near the mouth of Lee's River, on February 14, 1897." There is an adult taken at Chepachet, January 22, 1898, and a young bird taken at West Greenwich December 2, 1890 in the Smith collection.

(125) 337. **Buteo borealis (Gmel.).** RED-TAILED HAWK. *Hen Hawk.* — A not uncommon winter visitant, and uncommon summer resident. Lt. Wirt Robinson saw one at Newport, December 25, 1890, took others on January 21, 31, March 25, 1891. Mr. H. S. Hathaway writes that "a male was taken at Phillipsdale, on December 25, 1892, by Mr. Walter Barstow, a female at Pine Hill, Exeter, November 30, 1893, a male, in adult plumage, at Scituate on December 2 ?, 1893, and one in January, 1894." There are numerous other records.

(126) 339. **Buteo lineatus (Gmel.).** RED-SHOULDERED HAWK. *Hen Hawk.* — A common resident throughout the year.

---

[1] Osprey, Vol. I, No. 8, p. 111.

(127) 343. **Buteo latissimus (Wils.).** BROAD-WINGED HAWK.— An uncommon migrant, and rare summer resident. Nests have been taken at Gloucester, Johnston, Smithfield, and Kingston.
(April) to (October).

(128) 347a. **Archibuteo lagopus sancti johannis (Gmel.).** AMERICAN ROUGH-LEGGED HAWK.— A not uncommon migrant and winter visitant. Lt. Wirt Robinson saw one catch a rat on the beach near Fort Adams at Newport about October, 1887, he found two on April 12, 1888, shot by a farmer (Peckham) some weeks before, he saw one on January 16, 1889, two on February 24 and 27, 1889, three on March 14, one on March 22 and 26, 1889, one on December 16, 17 and 23, 1890, and two on January 6, 1891, all of these latter, including the two found dead, he writes "were seen in the valley that is now a part of the Newport golf grounds, attracted there by swarms of meadow mice." Mr. H. S. Hathaway writes that "a fine male was shot at Narragansett Pier, November 24, 1894, now in my collection, one in December, 1894, at Newport, and five were sent into Mr. J. M. Southwick's to be mounted in the fall of 1895."

(129) 349. **Aquila chrysaëtos (Linn.).** GOLDEN EAGLE. — Mr. J. M. Southwick recorded the capture of one "at Westerly, by Mr. J. B. Chapman," on February 17, 1887.[1] A young male was shot by Mr. Newton Dexter at Sakonnet in October, 1893,[2] and Mr. J. Hague had one alive in captivity for two or three years which, he stated, was taken in the State." An immature female was shot at Little Compton on December 13-15, 1898, by a Mr. Grinnell. Mr. F. T. Jencks writes that one was taken by Mr. Amasa Matheuson at Rockland (date unknown).

(130) 352. **Haliæetus leucocephalus (Linn.).** BALD EAGLE.— An irregular visitant. There was a pair reported to have remained the summer of 1882, near Pawtucket, and Mr.

---
[1] Southwick's List, p. 6.
[2] F. & S., Vol. XXVIII, No. 6, p. 106.

J. M. Southwick states that "during the past season (1887) a pair were located somewhere about the Seekonk River in East Providence." There are, however, no definite breeding records. It has been taken at Pawtucket, October 19, 1880, at Charlestown, December 8, 1894, at Block Island, July 25, 1897 at Charlestown Beach, November, 1886, at Warwick, 1862, and at Newport. There are numerous other records.*

(131) 354a. **Falco rusticolus gyrfalco (Linn.).** GYRFALCON. — One was taken near Providence by Mr. Newton Dexter in the winter of 1864-5, now in the Museum of Comparative Zoology collection, Cambridge, Mass.,[1] one at Point Judith, by Mr. J. S. Hopkins, October 11, 1883, now in the New England collection of the Boston Society of Natural History.[2]

(132) 353b. **Falco rusticolus obsoletus (Gmel.).** BLACK GYRFALCON. — A female was taken by Mr. A. O'D. Taylor on November 22, 1891, on Conanicut Island, now in the Newport Natural History Society collection,[3] and another female was taken by Mr. Arthur Scudder at Tiverton on December 26, 1896, now in the collection of Mr. A. C. Bent's of Taunton, Mass.[4] There is also a bird in the Smith collection taken at Newport, October 28, 1896.

(133) 356. **Falco peregrinus anatum (Bonap.).** DUCK HAWK. — A rare migrant. Mr. H. S. Hathaway writes that "a

---

* F. & S., Vol. XIX, No. 4, p. 65.

[1] Notes Rarer Birds Mass. J. A. Allen, Am. Nat. Vol. III, No. 10, p. 513, as *Falco sacer*; Baird, Brewer and Ridgway, His. No. Am. Birds, Vol. III, p. 115; in Coues and Stearns's New Eng. Bird Life, Vol. II, p. 111 as *F. gyrfalco islandicus*, in Samuel's Birds New Eng. and Ad. States, p. 576 as *F. sacer*; Allen's, Revised List Birds, Mass. Bull. Am Mus. Nat. His. Vol. 1, No. 7, p. 244. Revised Minot's Land and Game Birds, p. 479.

[2] Auk, Vol. I, No. I, p. 94; Ran. Notes, Vol. 1, No. 1, p. 6; O & O., Vol. 8, No. 12, p. 91. Allen's, Revised List Birds Mass. Bull. Am. Mus. Nat. His. Vol. I, No. 7, p. 244. Revised Minot's Land and Game Birds, p. 479.

[3] Auk, Vol. IX, No. 3, p. 300, 301, Revised Minot's Land and Game Birds, p. 480.

[4] Auk, Vol. XV, No. 1, p. 54.

Mr. Barstow shot one on the Seekonk River in 1881 or '82. The bird flew at his duck decoys; and that a young female, now in his collection, was shot at Newport on October 10, 1894, two days after a big northeast storm." Two were also taken at Point Judith,[1] and there are a number of other records without exact data. Mr. Newton Dexter writes "that they are not uncommon along shore in October." There is a female in the Smith collection taken on May 8, 1896, at Prudence Island.

(134) 357. **Falco columbarius Linn.** PIGEON HAWK. — A common migrant, especially in the fall.*
(April 5) to (May), September 3 to October 25.

(135) 360. **Falco sparverius (Linn.).** AMERICAN SPARROW HAWK. — An uncommon summer resident. Warwick, 1899.†
(February) to (November).

(136) 364. **Pandion haliaetus carolinensis (Gmel.).** AMERICAN OSPREY. *Fish Hawk.* — A common summer resident on Narragansett Bay. The birds at Bristol build their nests on cartwheels placed on poles by the farmers, as well as in trees,[2] and two nests, one on a flat steeple of a meeting house (Portsmouth, 1899)[3] and the other on a house chimney (Bristol, 1899) have been observed.
March 24 to October 21.

(137) 365. **Strix pratincola Bonap.** AMERICAN BARN OWL. — Of very rare occurrence. One was captured in November, 1886, by Mr. John Ryder (at Sand Pond) in Norwood, Warwick.[4] Col. J. H. Powel also includes this species in a list

---

[1] O. & O., Vol. 8, No. 12, p. 92.
* F. & S., Vol. XXXI, No. 15, p. 285.
† O. & O., Vol. 8, No. 8, p. 24.
[2] Auk, Vol. XII, No. 3, p. 300, and No. 4, p. 389.
Amer. Nat., Vol. IV, No. 1, p. 57.
Nid., Vol. I, No. 5, p. 72.
[3] Osprey, Vol. IV, No. 1, p. 13.
[4] Southwick's List, p. 7.
Ran. Notes, Vol. III, No. 12, p. 91.

sent to Mr. Sturtevant of birds taken near Newport. Mr. H. S. Hathaway writes that Mr. J. Baxter of Pawtucket shot one in Cumberland, and that a male was taken in December, 1891, now in the Brown University collection. There was also one taken by Mr. J. H. Tower at Charlestown on January 23, 1896.

(138) 366. **Asio wilsonianus (Less.).** AMERICAN LONG-EARED OWL. *Cat Owl.*— An uncommon resident. Mr. J. M. Southwick includes this species,[1] and Mr. H. S. Hathaway has a bird in his collection taken December 2, 1891. Mr. Sturtevant has an adult taken at Newport, March 21, 1899, by Mr. R. Pumpelly. Lt. Wirt Robinson writes that he shot one on October 2, 11, and November 13, 1888, at Newport.

(139) 367. **Asio accipitrinus (Pall.).** SHORT-EARED OWL. — An uncommon, but regular migrant. There is a questionable record of its breeding in late years. Mr. J. M. Southwick calls them less common than the preceding species, but we believe this not to be the case.[2] Mr. Hathaway writes that they were very plentiful in October, 1895, eight birds being brought in to Messrs. Southwick and Critchley to be mounted. Mr. Sturtevant has taken a number at Newport on the following dates: April 15 and August 14, 1896.[3] October 26 and November 18, 1890.
April, October 26 to December 23. A few undoubtedly winter.

(140) 368. **Syrnium nebulosum (Forst.).** BARRED OWL. *Hoot Owl.*— An uncommon resident, breeding regularly. Mr. J. M. Southwick calls it "one of our most common species." "During the winter of 1883 and 1884 they were wonderfully common and I was cognizant to the capture of more than fifty."[4] They are now, however, found in no such numbers.

---

[1] Southwick's List, p. 7.
[2] Southwick's List, p. 7.
[3] Auk, Vol. XIII, No. 3, p. 257.
[4] Southwick's List, p. 7.

(141) 370. **Scotiaptex cinereum** (Gmel.). GREAT GRAY OWL. *Spectral Owl.* — An extremely rare winter visitant. Mr. Newton Dexter writes that one was taken near Providence in 1870 and brought to Mr. John Hague, Taxidermist. Mr. J. M. Southwick records the capture of one in February 1883 (the correct date being March 25, 1883) at Fox Island, near Wickford. Mr. G. M. Gray purchased and mounted the bird and it is now in Mr. H. S. Hathaway's collection.[1] There was also one taken at Seekonk, Mass., just over the State line, about 1864, now in the Brown University collection.[2]

(142) 371. **Nyctala tengmalmi richardsoni** (Bonap.). RICHARDSON'S OWL. *Sparrow Owl.* — A very rare and irregular winter visitant. One was taken near Providence during the winter of 1880–81, a male was taken probably in Seekonk, Mass., December 18, 1882,[3] and one at Scituate during the winter of 1882–83, mounted by Mr. Daniel Seamans.[4] There is also a record of one brought to Mr. Hague, Taxidermist, by a young lady, taken in an old house in North Providence, January, 1881, now in Brown University collection.[5] A small Owl was captured by a farmer in a dog kennel on a Bristol farm and was kept in captivity a week or so. The bird finally died. Mr. Howe visited the farm a few days after and found the remains, the white beak and reddish plumage made its identification almost positive. Mr. H. S. Hathaway writes that he has one in his collection that was killed in the northern part of the State several years ago.

---

[1] Southwick's List, p. 9.
Ran. Notes, Vol. I, No. 7, p. 3.
Allen's, Revised List Birds Mass. Bull. Am. Mus. Nat. His. Vol. I, No. 7, p. 245.

[2] Bull. Nut. Orn. Club, Vol. VIII, No. 3, p. 183.

[3] Bull. Nut. Orn. Club, Vol. VI, No. 2, p. 123 and Vol. VIII, No. 2, p. 122.
Allen's Revised List Birds Mass. Bull. Am. Mus. Nat. His., Vol. I, No. 7, p. 245.
Coues and Stearns's, New Eng. Bird Life, Part II, p. 98.

[4] Ran. Notes, Vol. II, No. 5, p. 8.

[5] O. & O., Vol. 6, No. 2, p. 14.

(143) 372. **Nyctala acadica (Gmel.).** SAW-WHET OWL. *Acadian Owl.* — A not uncommon winter visitant. Mr. J. M. Southwick states that between January 8 and 21, 1887, he knew of the taking of several at Johnston, Seekonk, Mass., and at Arctic.[1] They have also been taken in Kingston[2] and Smithfield. (November) January 8 to 21. (February).

(144) 373. **Megascops asio (Linn.).** SCREECH OWL. *Cat Owl.* — A common resident. The most common of our Owls. December to (March).

(145) 375. **Bubo virginianus (Gmel.)** GREAT HORNED OWL. — A rare, irregular visitant, and rare summer resident. Nests have been taken at Albion, Washington County, and elsewhere. Mr. J. M. Southwick states that "they are not scarce," and that they breed, "proven by three fledglings taken from the old nest of a Fish Hawk, by Dr. Hall of Warren."[3] Mr. O. Durfee writes that "young birds have been brought up to Fall River from Tiverton."

(146) 376. **Nyctea nyctea (Linn.).** SNOWY OWL. *White Owl. Arctic Owl.* — An uncommon irregular winter visitant. They have been taken at Point Judith, Little Compton, Portsmouth, Tiverton, Bristol, Middletown, Providence,[4] Newport[4] Barrington, Warwick, and Pawtucket River.

(147) 377a. **Surnia ulula caparoch (Müll.).** AMERICAN HAWK OWL. — The only record is of one taken by Mr. W. A. Aldrich (exact locality not known.)

(148) 387. **Coccyzus americanus (Linn.).** YELLOW-BILLED CUCKOO. — A common summer resident, varying very perceptibly in numbers in different seasons.

---

[1] Southwick's List, p. 7.
[2] Ran. Notes, Vol. I, No. I, p. 4.
[3] Southwick's List, p. 8.
  Ran. Notes, Vol. I, No. 7, p. 3.
[4] O. & O., Vol. 8, No. 3, p. 24, and Vol. 10, No. 3, p. 48.

May 9 to September 25.   One record for October 23.

(149) 388. **Coccyzus erythrophthalmus** (Wils.). BLACK-BILLED CUCKOO. — A not uncommon summer resident.
May 11 to (September 25).

(150) 390. **Ceryle alcyon** (Linn.).   BELTED KINGFISHER. *Kingfisher.* — A common summer resident, and not uncommon in winter about open water.
April 2 to October 20.

(151) 393. **Dryobates villosus** (Linn.).   HAIRY WOODPECKER. — An uncommon winter visitant, and rare summer resident. Nests have been taken at Chepachet, Cranston, and elsewhere.
September 28 to (March).   Rarely in summer.

(152) 394c. **Dryobates pubescens medianus** (Swains.). DOWNY WOODPECKER. — A common resident, more abundant during the winter months.

(153) 402. **Sphyrapicus varius** (Linn.).   YELLOW-BELLIED SAPSUCKER. *Yellow-bellied Woodpecker.* — A not uncommon migrant.
March 23, April, October.

(154) 406. **Melanerpes erythrocephalus** (Linn.).   RED-HEADED WOODPECKER. — An irregular, but sometimes not uncommon migrant in the fall. It has been known to breed. Specimens have been taken at Prudence Island, Conanicut Island,[1] Newport, Warren, Providence, Point Judith, Oakland Beach, Warwick, where on July 28, 1882, a nest with young was found in a small oak.[2]

(155) 409. **Melanerpes carolinus** (Linn.).   RED-BELLIED WOODPECKER. — There are two birds, taken near Providence,

---
[1] Auk, Vol. XVI, No. 2, p. 189.
[2] F. & S., Vol. XIX, No. 4, p. 65.

in the New England collection of the Boston Society of Natural History.

(156) 412a. **Colaptes auratus luteus Bangs.** NORTHERN FLICKER. *Flicker. Yellow-hammer. Pigeon Woodpecker. Golden-winged Woodpecker. High Hole. Wake-up. Harry Wicket. Yellow Jay.* — An abundant resident. This species in this State does damage to buildings by making its holes under eaves and porches, and by pecking blinds etc., during the winter months.*

March 18 to November 14, common in winter.

(157) 417. **Antrostomus vociferus (Wils.).** WHIP-POOR-WILL. — A common summer resident. This species is almost entirely absent on many of the islands in Narragansett Bay and on Bristol promontory.

April 25 to (October).

(158) 420. **Chordeiles virginianus (Gmel.).** NIGHTHAWK. *Bull Bat.* — A rare migrant, (except in the northern and western portions of the State, where it is not uncommon), and summer resident. Dr. Rives also states that this species breeds.[1] Mr. O. Durfee writes that "until recently, if not at present it bred commonly on the islands of Rhode Island and Prudence." Mr. F. E. Newbury reports it breeds at Portsmouth and Mr. J. M. Southwick at Providence.

April 7 (May) September, October.

(159) 423. **Chaetura pelagica (Linn.).** CHIMNEY SWIFT. *Chimney Swallow.* — A common summer resident.*

April 26 to September 17.

(160) 428. **Trochilus colubris Linn.** RUBY-THROATED HUMMINGBIRD. — A common summer resident.

May 6 to September 23.

---

* Nid. Vol. II, No. 12, p. 170.
[1] Dr. Rives' List, p. 36.
* F. & S., Vol. 6, No. 17, p. 266.

(161) 444. **Tyrannus tyrannus (Linn.).** KINGBIRD.—An abundant summer resident.

May 1 to September 28.

(162) 452. **Myiarchus crinitus (Linn.).** CRESTED FLYCATCHER. *Great Crested Flycatcher.*— An uncommon summer resident. Formerly it nested regularly at Bristol, but of late years has become rare. In the northern portions of the State it seems however to have increased.

May 2 to (August).

(163) 456. **Sayornis phœbe (Lath.).** PHŒBE. *Bridge Pewee.*— A common summer resident of the northern and western portions of the State, but uncommon in southeastern portions.

March 28 to October 17.

(164) 459. **Contopus borealis (Swains.).** OLIVE-SIDED FLYCATCHER.— Mr. Eric Green took one on May 14, 1893 (exact locality not known). It is now in the Smith collection. Mr. F. T. Jencks reports having seen two in the spring of 1895.

(165) 461. **Contopus virens (Linn.).** WOOD PEWEE.— A common summer resident.

May 12 to September 24.

(166) 463. **Empidonax flaviventris Baird.** YELLOW-BELLIED FLYCATCHER.— An uncommon migrant. Mr. H. S. Hathaway writes "that Mr. C. H. Lawton took a pair on July 27?, 1885, in the extreme northeast end of the island of Rhode Island, and that another one was taken on August 7, 1887, and that Mr. George Gray informed him that he shot one at Centerdale."

(167) 467. **Empidonax minimus Baird.** LEAST FLYCATCHER.— A common summer resident, especially in all the towns and villages.

April 22 to (November).

(168) 474. **Otocoris alpestris (Linn.).** HORNED LARK. *Shore Lark.* — A common migrant and winter resident.*
September 25 to April 15.

(169) 474b. **Otocoris alpestris praticola Hensh.** PRAIRIE HORNED LARK. — There is one bird in the Smith collection taken at Pawtuxet, November 25, 1889, by Mr. J. W. Staintor.

(170) 477. **Cyanocitta cristata (Linn.).** BLUE JAY. — An abundant resident throughout the wooded portions of the State.

(171) 488. **Corvus americanus Aud.** AMERICAN CROW. *Crow.* — An abundant resident. During the fall and winter months there is an interesting daily migration at sunrise and sunset to and from their roosts. (See Migration.)

(172) 494. **Dolichonyx oryzivorus (Linn.).** BOBOLINK. *Reed Bird.* — An abundant summer resident.†
May 10 to September 25.

(173) 495. **Molothrus ater (Bodd.).** COWBIRD. *Lazy-bird. Cow Blackbird. Cow Bunting.* — An abundant summer resident. One winter record of a pair taken January 19, 1887.[1]
March 14 to November 3.

(174) 498. **Agelaius phœniceus (Linn.).** RED-WINGED BLACKBIRD. *Red-winger. Quonk-a-ree. Red and Buff Shouldered Blackbird.* — An abundant summer resident. There is a young male taken at Cranston, on December 15, 1894, now in the Smith collection.‡
March 7 to October 13. One record for early arrival, February 10 to 13.

---

* F. & S., Vol. XXIV, No. 12, p. 225.
† F. & S., Vol. 6, No. 17, p. 266.
[1] Southwick's List, p. 10.
‡ Ran. Notes, Vol. III, No. 4, p. 27.
F. & S., Vol. XXII, No. 9, p. 165.

(175) 501. **Sturnella magna** (Linn.). MEADOWLARK. *Marsh Quail.* —An abundant resident. This species winters in large numbers, flocks of over one hundred birds often inhabiting a salt marsh, which they prefer to meadows in winter, on account of their remaining open. They sleep at night in the long salt grass huddled together, and when disturbed rise in a body and fly wildly about.

(176) 506. **Icterus spurius** (Linn.). ORCHARD ORIOLE. — A rare summer resident, perhaps increasing in numbers. Nests have been found at Warwick Neck, Barrington, Newport and elsewhere. They were undoubtedly more common in past years, they certainly were so at Bristol. Mr. J. S. Howland records their arrival at Newport from May 14 to 19, in 1876, as if they were somewhat regular and common in their occurrence.[1]
May 14 to (July).

(177) 507. **Icterus galbula** (Linn.). BALTIMORE ORIOLE. *Firebird.* — A common summer resident.*
May 4 to August 28.

(178) 509. **Scolecophagus carolinus** (Müll.). RUSTY GRACKLE. — A not uncommon migrant, though perhaps less so than in Massachusetts.
March 13 to June 5. (September) to October 23.

(179) 511. **Quiscalus quiscula** (Linn.). PURPLE GRACKLE. *Crow Blackbird.* — A common summer resident in the southern portions of the State. A series of specimens taken at Middletown, Newport and Bristol were typical *quiscula* with but few intermediates between the two races. Specimens from Providence, beside a few intermediates, and one from North Smithfield were typical *æneus*. This State therefore seems to be on the border line between the two races.†

---

[1] F. & S., Vol. 6, No. 17, p. 266.
* F. & S., Vol. 6, No. 17, p. 266.
† F. & S., Vol. XXII, No. 9, p. 165.
Nid., Vol. II, No. 12, p. 170.

March 6 to September 21 (November).

(180) 511b. **Quiscalus quiscula æneus (Ridgw.). Bronzed Grackle.** — A common summer resident in the northern portions of the State. (See preceding species).
March 5 to November 1.

(181) 515. **Pinicola enucleator canadensis (Cab.) Canadian Pine Grosbeak.** *Pine Grosbeak.* — A rare and irregular winter visitant. Mr. Newton Dexter writes that it was very abundant in winter of 1853. Dr. Rives states that this species visited Rhode Island during the winter of 1863–4, as it did all New England. Only a few full plumaged males were seen, most of the birds being females and young.[1] Mr. F. T. Jencks writes that "during the winter of 1872–3 and again ten years later they were quite plentiful." During the winter of 1892–3, the year of the remarkable flight, they visited the State in large numbers.[2] Mr. H. S. Hathaway writes " that a few flocks were seen at Smithfield during the week of February 2–8, 1896, a large portion of them being in the red plumage." Mr. O. Durfee says "that these birds rarely leave the pine timber of Taunton to come down the river."[*]

(182) 517. **Carpodacus purpureus (Gmel.). Purple Finch.** *Red Linnet.* — A common summer resident, and frequently seen during the winter months.
April 14 to October. Winter.

(183) **Passer domesticus (Linn.). House Sparrow.** *English Sparrow.* — An abundant resident. Not only is it an inhabitant of the towns and villages, but also of the farms. It even builds its nest under the Osprey's, among the larger sticks of the foundation.[†]

---

[1] Dr. Rives' List, p. 32.
[2] Auk, Vol. XII, No. 3, p. 254.
[*] F. & S., Vol. XXII, No. 5, p. 83.
Southwick's List, p. 9.
[†] Am. Nat., Vol. VIII, No. 11, p. 692.

(184) 521. **Loxia curvirostra minor (Brehm).** AMERICAN CROSSBILL. *Red Crossbill.* — A common, but irregular visitant, generally in winter and spring.*

January, February, March, April, May, and December.

(185) 522. **Loxia leucoptera Gmel.** WHITE-WINGED CROSSBILL. — An extremely rare and irregular visitant in winter. Southwick includes it in his List,[1] and Mr. H. S. Hathaway writes that it is rare. Mr. F. T. Jencks took a pair just east of the Rhode Island line in Seekonk, Mass., in the winter of 1874 or 1875.

(186) 528. **Acanthis linaria (Linn.).** REDPOLL. — An irregular winter visitant, during some seasons common. Mr. F. T. Jencks writes that flocks of hundreds were seen in the winter of 1874–75. Mr. H. S. Hathaway writes "that he saw a flock of over one hundred, March 25, 1880." Lt. Wirt Robinson took two at Newport on March 14, 1888. Mr. F. E. Newbury saw a small flock near Greenville on March 7, 1897. Undoubtedly the Greater Redpoll (*A. linaria rostrata*) also occurs.

(October) to (April).

(187) 529. **Astragalinus tristis (Linn.).** AMERICAN GOLDFINCH. — *Summer Yellowbird. Yellowbird. Thistlebird.* — An abundant summer, and common winter resident.

(188) 533. **Spinus pinus (Wils.).** PINE SISKIN. *Gray Linnet.* — An irregular winter visitant, often common.†

October to April 17.

(189) 534. **Passerina nivalis (Linn.).** SNOWFLAKE. *Snow Bunting.* — A common winter resident, abundant on migrations. They appear and leave inland, with heavy snows.‡

November 25 to March 29.

---

* Ran. Notes, Vol. I, No. 6, p. 6.
[1] Southwick's List, p. 10.
† F. & S., Vol. XXII, No. 5, p. 83.
Coues and Stearns's, "New Eng. Bird Life," Vol. I, p. 229.
‡ F. & S., Vol. XXIV, No. 12, p. 225.

(190) 536. **Calcarius lapponicus (Linn.).** LAPLAND LONG-SPUR. — A probably not uncommon migrant, of which we have, however, but a few records. It has been reported by Mr. C. H. Lawton from Newport, and Mr. H. S. Hathaway writes that he bought an adult bird shot at Warwick, on January 6, 1891. A male was also taken at Gaspee Point, February 16, 1899, now in the Smith collection.

(191) 540. **Pocecetes gramineus (Gmel.).** VESPER SPARROW. *Grass Finch.* — A common summer resident. During the summer of 1899, however, very rare on account of the blizzard in the South in February.

March 19 to October 17.

(192) 541. **Ammodramus princeps (Mayn.).** IPSWICH SPARROW. — A common migrant, and not uncommon winter resident[1] on the sand dunes along the coast.*

(October 20) November 6 to April 11.

(193) 542a. **Ammodramus sandwichensis savanna (Wils.).** SAVANNA SPARROW. — An abundant migrant, and not uncommon summer resident, especially near salt water.†

April 4 to October 20.

(194) 546. **Ammodramus savannarum passerinus (Wils.).** GRASSHOPPER SPARROW. *Yellow-winged Sparrow.* — A not uncommon summer resident, though somewhat local in its distribution.

May 8 to October 8.

(195) 547. **Ammodramus henslowii (Aud.).** HENSLOW'S SPARROW. — Mr. F. T. Jencks took a male the last of April, 1874, in Cranston. Lt. Wirt Robinson's record, he writes,

---

[1] Auk, Vol. XVI, No. 2, p. 89.
* Ran. Notes, Vol. II, No. 3, p. 17.
† F. & S., Vol. XV, No. 17, p. 266.

should be expunged, he having made a mistake in identification at the time.¹

(196) 549. **Ammodramus caudacutus (Gmel.).** SHARP-TAILED SPARROW. — A common summer resident in all the salt marshes.*

(May) to October 2.

(197) 549b. **Ammodramus caudacutus subvirgatus Dwight.** ACADIAN SHARP-TAILED SPARROW. — Undoubtedly not an uncommon migrant with *A. nelsoni*, but we have only one record of a bird taken at Point Judith on the very early date of April 27, 1887.²

April 27 (September, October).

(198) 550. **Ammodramus maritimus (Wils.).** SEASIDE SPARROW. — A not uncommon summer resident at Point Judith marshes. It also has been taken in the Middletown marshes, in spring and summer.³†

(199) 554. **Zonotrichia leucophrys (Forst.).** WHITE-CROWNED SPARROW. — An uncommon migrant. Mr. F. T. Jencks shot one at Drownville in the spring of 1891 and also took one in the fall, and has seen others. Lt. Robinson shot one at Newport, October 9, 1888, and on October 11, 1889, and also saw several on October 12, 1889. One was taken by Mr. Erik Green in Cat Swamp, Providence, on July 9, now in the Smith collection.

(May), October 12.

(200) 558. **Zonotrichia albicollis (Gmel.).** WHITE-THROATED SPARROW. — A common migrant. It has wintered at Drownville.‡

---

[1] Auk, Vol. VI, No. 2, p. 194 and Vol. XVI, No. 4, p. 356.
* Am. Nat., Vol. III, No. 4, p. 229.
[2] Auk, Vol. IV, No. 2, p. 136.
[3] Auk, Vol. XVI, No. 2, p. 229 and No. 2, p. 322.
† Am. Nat., Vol. III, No. 4, p. 229.
‡ F. & S., Vol. XXIV, No. 12, p. 225.

March 29 to April 28 (May 10), (October 1) to November 1.

(201) 559. **Spizella monticola (Gmel.).** TREE SPARROW.
— An abundant winter resident.
November 14 to April 11.

(202) 560. **Spizella socialis (Wils.).** CHIPPING SPARROW.
Chippy. — An abundant summer resident.
April (4?) 20 to October 17.

(203) 563. **Spizella pusilla (Wils.).** FIELD SPARROW.
Ground Sparrow. — A common summer resident. Mr. Eli W. Blake took one at East Providence in January, 1886, and Mr. James Bilson found one dead in Roger Williams Park, Providence, on February 6, now in the Smith collection.
March 5, April to October 23. January. February.

(204) 567. **Junco hyemalis (Linn.).** SLATE-COLORED JUNCO.
Junco. Snowbird. — A common winter resident.
October 2 to April 4, May 19.

(205) 581. **Melopiza fasciata (Gmel.).** SONG SPARROW.—
An abundant summer resident, and not uncommon winter resident.*
March 1 to November 14. Winter.

(206) 583. **Melopiza lincolni (Aud.).** LINCOLN'S SPARROW. *Lincoln's Finch.* — An uncommon migrant. Mr. F. B. Webster took one in 1880 on Bucklin's Island, Pawtucket River, now in the Smith collection, and Mr. Howard Mason took one at Cranston in the fall of 1897. Mr. Hathaway took a male on September 28, 1898, in Warwick.[1]
(May), September.

(207) 584. **Melospiza georgiana (Lath.).** SWAMP SPAR-

---

* Ran. Notes, Vol. III, No. 4, p. 27.
F. & S., Vol. VI, No. 17, p. 266.
[1] Osprey, Vol. III, No. 7, p. 110.

row.— A common summer resident, somewhat locally distributed. Very abundant during fall migration.

April 4 to November 14.

(208) 585. **Passerella iliaca (Merr.).** Fox Sparrow.— A common migrant. Mr. H. S. Hathaway writes "that one was killed at Johnston May 21, 1891," a late date. A bird wintered in Cranston in 1899, near Mr. Hathaway's house.[1]

March 19 to April 20, October 13 to November 14.

(209) 587. **Pipilo erythrophthalmus (Linn.).** Towhee. *Chewink.*— A common summer resident. Mr. W. W. Bull shot a female at Newport on January 14, 1896.

April 19 to October 13.

(210) 595. **Zamelodia ludoviciana (Linn.).** Rose-breasted Grosbeak. — A common summer resident in the northern portions of the State, but rare in the southeastern portions.

May 5 to October 2.

(211) 598. **Cyanospiza cyanea (Linn.).** Indigo Bunting. *Indigobird.*— A common summer resident, like the foregoing species, of the northern and western portions of the State, absent or rare in the southeastern portions.

(May 10) to (September 25).

(212) 601. **Cyanospiza ciris (Linn.).** Painted Bunting. — *Nonpareil.* An accidental visitant, or escaped cage bird. Mr. Daniel Seamans took one at Scituate in the summer of 1882.[2]

(213) 604. **Spiza americana (Gmel.).** Dickcissel. *Black-throated Bunting.*— Lt. Wirt Robinson shot a young bird at Newport, September 25, 1888. The bird, he writes, he shot by mistake, when he was after some Bobolinks, in a cornfield on

---

[1] Osprey, Vol. III, No. 7, p. 111.
[2] Ran Notes, Vol. II, No. 5, p. 8.

Peckham's place. It was perched on a corn tassel. Dr. W. C. Rives now has the skin. This is the only record for this species.[1]

(214) 608. **Piranga erythromelas Vieill.** SCARLET TANAGER. — A common summer resident of all but the southeastern portions of the State, where it is uncommon. A male, in full spring plumage, was taken at Apponaug on April 4, 1891.[2]
May 4 to October 9.

(215) 610. **Piranga rubra (Linn.).** SUMMER TANAGER. Mr. Newton Dexter reports two seen in Providence, and one was taken on the Ten Mile River by Mr. C. M. Carpenter,[3] now in the Smith collection.

(216) 611. **Progne subis (Linn.).** PURPLE MARTIN. — An uncommon summer resident, and common migrant. Formerly much more abundant.[*]
April 25 to September 16.

(217) 612. **Petrochelidon lunifrons (Say).** CLIFF SWALLOW. *Eave Swallow.* — An uncommon summer resident, becoming rarer each year.
(May) to August 28.

(218) 613. **Hirundo erythrogastra (Bodd.).** BARN SWALLOW. — An abundant summer resident. Mr. T. M. Brewer in an article entitled " Sea-side Ornithology " published in 1870, writes "yet we can remember when the rocks of Newport and Nahant were their primitive breeding ground."[4] At the present

---

[1] Auk, Vol. VI, No. 2, p. 194.
NOTE: There was a South American Finch (*Gubernatrix cristatella*) taken near Providence by Rev. C. H. Baggs on July 7, 1880. Bull. Nut. Orn. Club, Vol. V, No. 4, p. 240. Coues and Stearns's New Eng. Bird Life, Part I, p. 29.
[2] O. & O., Vol. 16, No. 5, p. 78.
[3] Bull. Nut. Orn. Club, Vol. II, No. 1, p. 21.
Birds of Conn. Merriam, Trans. Conn. Acad. Vol. IV, p. 21.
Coues and Stearns's, New Eng. Bird Life, Part I, p. 180.
[*] F. & S., Vol. 6, No. 17, p. 266.
[4] Am. Nat., Vol. III, No. 4, p. 228.

day this species still nests where it has for years in the chasm called "Purgatory" at the westerly end of the Second Beach, Middletown (see Frontispiece); building their nests in the little inaccessible crevices of the vertical walls, where they usually select a spot that is protected from rain by a projecting bit of rock. No Cliff Swallows (*Petrochelidon lunifrons*) breed in this chasm, a place apparently more suited to them, than to the Barn Swallows.

April 10 to September 17.

(219) 614. **Tachycineta bicolor (Vieill.).** TREE SWALLOW. *White-breasted Swallow.* — A common summer resident, and abundant migrant. One winter record.[1] *

March 12, March 28 to October 17.

(220) 616. **Clivicola riparia (Linn.).** BANK SWALLOW. — A common local summer resident (see Illustration). Colonies at Sachuest Point, Conanicut Island and elsewhere.

April 30 to August 29.

(221) 619. **Ampelis cedrorum (Vieill.).** CEDAR WAXWING. *Cedarbird.* — A common summer resident. Occasionally seen during the winter months.

February 1 to September 27. January.

(222) 621. **Lanius borealis Vieill.** NORTHERN SHRIKE. *Butcherbird.* — A not uncommon winter visitant, but varying in numbers in different seasons.

(November 1) to March 23.

(223) 622a. **Lanius ludovicianus excubitorides (Swains.).** WHITE-RUMPED SHRIKE. — A rare autumn and winter visitant. Mr. F. T. Jencks took a bird in Cranston on September 2, 1873,[2]

---

[1] F. & S., Vol. XVII, No. 9, p. 203.
* F. & S., Vol. 6, No. 17, p. 266 and Vol. XXII, No. 9, p. 165.
[2] Bull. Nut. Orn. Club, Vol. II, No. 1, p. 21
Coues and Stearns's New Eng. Bird Life. Part I, p. 212.

and Mr. LeRoy King shot one at Newport on August 29, 1898, which is now in his collection.[1]

(224) 624. **Vireo olivaceus (Linn.).** RED-EYED VIREO. — An abundant summer resident.
April 29 to September 24.

(225) 626. **Vireo philadelphicus (Cass.).** — PHILADELPHIA VIREO. — A very rare migrant. One was shot by Mr. Kristian Hansen at Drownville on May 31, 1891, recorded by Mr. F. T. Jencks, it is now in the Smith collection.[2]

(226) 627. **Vireo gilvus (Vieill.).** WARBLING VIREO. A common summer resident, especially in cities and villages.*
May 5 to (September 25.)

(227) 628. **Vireo flavifrons Vieill.** YELLOW-THROATED VIREO. — A not uncommon summer resident.
May 1 to (September).

(228) 629. **Vireo solitarius (Wils.).** BLUE-HEADED VIREO. *Solitary Vireo.* — An uncommon summer resident, (Washington), and common migrant.
April 23 to October 22.

(229) 631. **Vireo noveboracensis (Gmel.).** WHITE-EYED VIREO. — A rather common summer resident, though somewhat local.†
May 7 to September 18.

(230) 636. **Mniotilta varia (Linn.).** BLACK AND WHITE WARBLER. *Black and White Creeper. Black and White Creeping Warbler.* — A common summer resident.
April 15 to September 24.

---

[1] Auk, Vol. XVI, No. 2, p. 190.
[2] Collector's Monthly, Conn. Vol. 2, No. 12, p. 72.
* F. & S., Vol. 6, No. 17, p. 266.
† O. & O., Vol. 9, No. 5, p. 58.

(231) 637. **Protonotaria citrea (Bodd.).** PROTHONOTARY WARBLER. *Golden Swamp Warbler.* — A male was shot on April 20 or 21, 1884, at South Kingston by Mr. Herbert Holland, and recorded by Mr. R. G. Hazard 2nd,[1] and Mr. H. S. Hathaway writes that Mr. William Deardon shot one at Lonsdale, April 29, 1892, now in the Smith collection, and a male in the same locality on April 19, 1893.

(232) 641. **Helminthophila pinus (Linn.).** BLUE-WINGED WARBLER. — A very rare or accidental summer resident. A nest was taken at Gloucester by Mr. C. E. Doe, on May 30, 1890, on the authority of Mr. F. E. Newbury.

(233) 642. **Helminthophila chrysoptera (Linn.).** GOLDEN-WINGED WARBLER. — A very rare summer resident. Mr. F. T. Jencks saw one in Providence in May, 1880.

May to (August 25).

(234) 645. **Helminthophila rubricapilla (Wils.).** NASHVILLE WARBLER. — An uncommon summer resident, but not uncommon migrant.

May 4 to (October).

(235) 646. **Helminthophila celata (Say.).** ORANGE-CROWNED WARBLER. A very rare migrant. One was shot by Mr. F. T. Jencks at Cranston, December 3, 1874.[2] A male was taken in East Providence on May 9, 1891.

(236) 647. **Helminthophila peregrina (Wils.).** TENNESSEE WARBLER. — A rare migrant. A male was taken at Centredale by Mr. Walter Angell on September 18, 1886,[3] now in the

---

[1] Auk, Vol. I, No. 3, p. 290.
Ran. Notes, Vol. I, No. 6, p. 3, and Vol. II, No. 5, p. 8.
Allen's Revised List Birds Mass. Bull. Am. Mus. Nat. His. Vol. I, p. 255.
[2] Bull. Nut. Orn. Club, Vol. II, No. 1, p. 121.
Allen's Revised List Birds Mass. Bull. Am. Mus. Nat. His. Vol. I, p. 256.
Coues and Stearns's New Eng. Bird Life. Part I, p. 119.
[3] Ran. Notes, Vol. III, No. 10, p. 79.

Smith collection,[1] and Mr. H. S. Hathaway took a male at Warwick on May 18, 1898.[1]

May 18 to September 18.

(237) 648. **Compsothlypis americana usneæ Brewster.** NORTHERN PARULA WARBLER. — A common migrant, and local summer resident, breeding in the southern portions of the State. Kingston, Tiverton, Mount Hope, and elsewhere.

April 30 to October 1.

(238) 650. **Dendroica tigrina (Gmel.).** CAPE MAY WARBLER. — A rare migrant. Mr. F. T. Jencks shot one in Cranston about 1879. A male was shot at Lonsdale, May 14, 1890, which is now in the Brown University collection. Another male was also taken in Lonsdale between May 20 and 25, 1890, now in Mr. H. S. Hathaway's collection. There is also one in the Smith collection.

May 13 to (25) (August 25 to September 15).

(239) 652. **Dendroica aestiva (Gmel.).** YELLOW WARBLER. *Summer Yellowbird. Wild Canary. Cotton Wren.* A common summer resident.

April 17, May 1 to (September 30).

(240) 654. **Dendroica cærulescens (Gmel.).** BLACK-THROATED BLUE WARBLER. — A not uncommon migrant.

May 14 to 22 (September 25) to October 17.

(241) 655. **Dendroica coronata (Linn.).** MYRTLE WARBLER. *Yellow-rumped Warbler. Golden-crowned Warbler.* — A common winter resident, and abundant migrant.

September 24 to April 20.

(242) 657. **Dendroica maculosa (Gmel.).** MAGNOLIA WARBLER. — A not uncommon migrant.

May 14 to (25) (September 14) to October 8.

---

[1] Osprey, Vol. III, No. 7. p. 110.

(243) 658. **Dendroica rara Wils.** CERULEAN WARBLER. — A very rare, or accidental migrant. One was reported by Mr. Ruthven Deane, a male, taken by Mr. C. M. Carpenter near Cumberland Hill, May 22, 1878.[1] A male was taken by Mr. William Deardon on May 14, 1893, at Lonsdale.

(244) 659. **Dendroica pensylvanica (Linn.).** CHESTNUT-SIDED WARBLER. — A common summer resident.
May 1 to (September 15).

(245) 660. **Dendroica castanea (Wils.).** BAY-BREASTED WARBLER. — An uncommon migrant.
May 9 to (20). (September 15 to 30).

(246) 661. **Dendroica striata (Forst.).** BLACK-POLL WARBLER. — A common migrant in the spring; very abundant in fall.
May 10 to 31, September 23 to October 23.

(247) 662. **Dendroica blackburniæ (Gmel.).** BLACKBURNIAN WARBLER. — An uncommon migrant.
May 14 to (20) (September 17 to October).

(248) 667. **Dendroica virens (Gmel.).** BLACK-THROATED GREEN WARBLER. — A common summer resident.
April 25 to October 13.

(249) 671. **Dendroica vigorsii (Aud.).** PINE WARBLER. *Pine-creeping Warbler.* — A common summer resident wherever there is pitch pine growth. Lt. Wirt Robinson reports one at Newport on November 15, 1890.
April 9 to October 29.

(250) 672. **Dendroica palmarum (Gmel.).** PALM WARBLER. — A rare fall migrant. Mr. H. S. Hathaway took one at Brightman's Pond, Westerly, September 21, 1896, now in the Smith collection.

---

[1] Bull. Nut. Orn. Club, Vol. IV, No. 3, p. 185.
Coues and Stearns's New Eng. Bird Life. Part I, p. 130.

(251) 672a. **Dendroica palmarum hypochrysea** Ridgw. YELLOW PALM WARBLER. — A common spring, but uncommon fall migrant.

April 5 to 22 (October 1 to 15).

(252) 673. **Dendroica discolor** (Vieill.) PRAIRIE WARBLER. — A common, local summer resident, nesting in bay bushes. May 4 to (September 20).

(253) 674. **Seiurus aurocapillus** (Linn.). OVEN-BIRD. — An abundant summer resident.

May 2 to September 28.

(254) 675. **Seiurus noveboracensis** (Gmel.). WATER-THRUSH. — A not uncommon migrant.

May 9 to (20), (August 15 to October 15).

(255) 676. **Seiurus motacilla** (Vieill.) LOUISIANA WATER-THRUSH. *Large-billed Water-Thrush.* — A rare summer resident. Mr. F. T. Jencks writes "probably generally distributed in the wilder portions of the southwestern section of the State." Mr. Ruthven Deane reports that Mr. F. T. Jencks took a pair on May 11, 1877, at Johnston; on May 2, 1879, in West Greenwich he took two males. About the middle of the same month he found a pair at the same locality, and another pair at a point some four miles distant. He also secured one at Point Judith.[1] A female was also taken in West Greenwich on May 17, 1887.

(256) 677. **Geothlypis formosa** (Wils.). KENTUCKY WARBLER. — Lt. Wirt Robinson writes that he observed a bird near Fort Adams, Newport, in the spring of 1890. Although he was unable to secure the bird, he is practically sure of its identity.

(257) 678. **Geothlypis agilis** (Wils.). CONNECTICUT

---

[1] Bull. Nut. Orn. Club, Vol. V, No. 2, p. 116.

O. & O., Vol. 7, No. 15, p. 114.

Allen's Revised List Birds of Mass. Bull. Am. Mus. Nat. His. Vol. I, p. 258.

Coues and Stearns's New Eng. Bird Life. Part I, p. 159.

WARBLER. — An uncommon autumn migrant. Mr. H. S. Hathaway writes that a young male was shot in Warwick on November 12, 1898.*

September 24 to October 2.

(258) 679. **Geothlypis philadelphia (Wils.).** MOURNING WARBLER. — A rare migrant. Has been taken at Warwick, Cranston, Pawtuxet, and elsewhere.[1] There is one in the Smith collection.

May 21 to June 5 (September 12 to 30).

(259) 681. **Geothlypis trichas (Linn.).** MARYLAND YELLOW-THROAT. — An abundant summer resident.

April 22 to October 17. Possibly winters.

(260) 683. **Icteria virens (Linn.).** YELLOW-BREASTED CHAT. — A locally common summer resident in the southern portions of the State. It has shown a perceptible increase in numbers during the past twenty years.

May 3 to ———.

(261) 684. **Wilsonia mitrata (Gmel.).** HOODED WARBLER. — A female, taken at Kingston, is in the New England collection of the Boston Society of Natural History.

(262) 685. **Wilsonia pusilla (Wils.).** WILSON'S WARBLER. — An uncommon migrant. There is a record for November 30, 1882, by Mr. C. M. Carpenter at French Camp. Mr. F. P. Drowne, took one in North Providence, May 15, 1897, now in the Smith collection.

May 15 to 22 (September 1 to 25).

(263) 686. **Wilsonia canadensis (Linn.).** CANADIAN WARBLER. — An uncommon migrant, and summer resident. Mr.

---

* Ran. Notes, Vol. I, No. 11, p. 7.
[1] Osprey, Vol. III, No. 7, p. 110

J. H. Sage reports them breeding near Noyes' Beach, and Mr. F. T. Jencks in June at Johnston.*

May (16) to 20 (September 5 to 25).

(264) 687. **Setophaga ruticilla** (Linn.). AMERICAN REDSTART. — A common summer resident.

May 1 to (October 5).

(265) 697. **Anthus pensilvanicus** (Lath.). AMERICAN PIPIT. — A common migrant.†

September 25 to October 16, March 29 to (May 15).

(267) 703. **Mimus polyglottos** (Linn.). MOCKINGBIRD. — A very rare summer resident. Mr. N. W. Thatcher took a bird in East Providence in 1877. Mr. Harry G. White reported one singing at Newport on November 2, 1888.[1] Mr. F. T. Jencks observed one at Drownville, October 18, 1891,[2] and he has since seen two others there. Lt. Wirt Robinson writes that he saw one at Newport on November 5 and 12, 1889. There was a pair in Roger Williams Park, Providence, in the autumn of 1897. How many of the Mockingbirds recorded are escaped cage birds it is impossible to state, the species no doubt occurs, however, in its wild state. ‡

(March) to November 2.

(267) 704. **Galeoscoptes carolinensis** (Linn.). CATBIRD. — An abundant summer resident. There is one very early record, March 19, 1897, at Bristol. The bird may have wintered.§

April 24 to October 2.

(268) 705. **Harporhynchus rufus** (Linn.). BROWN THRASHER. *Brown Thrush.* — A common summer resident.

---

* Osprey, Vol. III, No. 7, p. 110.
† F. & S., Vol. XXIV, No. 12, p. 225.
[1] O. & O., Vol. 13, No. 12, p. 192.
[2] Collector's Monthly, Conn. Vol. 2, No. 12, p. 73.
‡ Coues and Stearns's, New Eng. Bird Life, Part I, p. 62.
§ F. & S., Vol. 6, No. 17, p. 266.

There is a record for January 30, 1886, at Johnston, and a doubtful one for February, 1882, at Pawtucket.[1][*]

April 14[2] to September 29 (October 20). January.

(269) 718. **Thryothorus ludovicianus (Lath.).** CAROLINA WREN. — A very rare summer resident. A male was taken by Mr. G. M. Gray on August 14, 1880, at Bristol.[3] There is also an interesting record for this species at Peacedale, summer and autumn of 1898.[4] Mr. F. T. Jencks reports having seen one near his house in Barrington for three successive summers. Last year they were not noted but a male arrived on April 9, 1899. They have occurred usually up to October, and have probably bred. Mr. Sturtevant took a male, and three young just able to fly on May 11, 1899, at Middletown; the young on account of their age could not have been far from their nest. This is the first actual breeding record for New England.[5]

April 9 to November 28.

(270) 721. **Troglodytes aëdon Vieill.** HOUSE WREN. — Formerly a common summer resident, but now only locally common.

April 26 to (September 25).

(271) 722. **Anorthura hiemalis Vieill.** WINTER WREN. — A not uncommon fall migrant, and rare winter resident. Mr. H. S. Hathaway and Mr. F. T. Jencks record the wintering of this species near their homes in Cranston and Drownville, winter of 1898-9.[6]

(September 25) to November 14 (April 5) to (May).

---

[1] O. & O., Vol. II, No. 6, p. 84.
[*] F. & S., Vol. 6, No. 17, p. 266.
[2] Ran. Notes, Vol. III, No. 5, p. 37.
[3] Bull. Nut. Orn. Club, Vol. V, No. 4, p. 237.
Allen's Revised List Birds Mass. Bull. Am. Mus. Nat. His. Vol. 1, p. 260.
[4] Auk, Vol. XVI, No. 1, p. 83 and [5] No. 3, p. 284.
[6] Osprey, Vol. III, No. 7, p. 111.

ANNOTATED LIST.

(272) 724. **Cistothorus stellaris (Licht.).** SHORT-BILLED MARSH WREN.— A rare summer resident. Mr. B. LaFarge found it nesting at Newport.[1]

(May 15) to (October 1).

(273) 725. **Cistothorus palustris (Wils.).** LONG-BILLED MARSH WREN.— An abundant local summer resident. It is found in the Newport, Middletown and Point Judith marshes. They can be heard singing all night throughout the summer.

May (15) 30 to October 6, a few possibly winter.

(274) 726. **Certhia familiaris fusca (Barton).** BROWN CREEPER.— A common migrant, and winter resident.

September 2 to April 26.

(275) 727. **Sitta carolinensis Lath.** WHITE-BREASTED NUTHATCH.— A not uncommon migrant, and winter resident, breeding locally.

September 14 to April 22, a few in summer.

(276) 728. **Sitta canadensis Linn.** RED-BREASTED NUTHATCH.— A not uncommon spring, and fall migrant, and winter resident.

September 9 to October 3 (November 25).

(277) 735. **Parus atricapillus (Linn.).** CHICKADEE. *Black-capped Titmouse.*— An abundant resident, observed more often, however, during the winter months.

(278) 740. **Parus hudsonicus Forst.** HUDSONIAN CHICKADEE.— A casual winter visitant. A bird was taken at Smithfield on November 1, 1880, by Mr. Thomas Adcock, and recorded by Mr. F. T. Jencks.[2]

---

[1] Dr. Rives' List, p. 35.
[2] Bull. Nut. Orn. Club, Vol. VI, No.— p. 54.
Allen's Revised List Birds Mass. Bull. Am. Mus. Nat. Hist., Vol. I, p. 261.

(279) 748. **Regulus satrapa Licht.** GOLDEN-CROWNED KINGLET. *Golden-crested Wren.* — A common migrant, and winter resident.

October 12 to April 2.

(280) 749. **Regulus calendula (Linn.).** RUBY-CROWNED KINGLET. — A common migrant.

April (10) 25 to May 6 (October 10 to November 5).

(281) 751. **Poliotila cærulea (Linn.).** BLUE-GRAY GNATCATCHER. — A casual visitant. Mr. H. A. Purdie reported three or four seen by Mr. F. T. Jencks in Providence, May 23, 1875.[1] A male was shot by a Mr. E. I. Shores, at Silver Spring, June 24, 1875. Mr. C. M. Carpenter has shot one or more in the State.

(282) 755. **Hylocichla mustelinus Gmel.** WOOD THRUSH. — A common summer resident in the wooded portions of the State.

May 4 to (September 15).

(283) 756. **Hylocichla fuscescens Steph.** WILSON'S THRUSH. *Veery.* — An abundant summer resident.

April 29 to (September 8).

(284) 756a. **Hylocichla fuscescens salicicola (Ridgw.).** WILLOW THRUSH. — A bird was taken on September 25, 1885, by Mr. R. L. Agassiz at Newport, and a male at Bristol, on September 24, 1899, by Mr. Howe.

(285) 757. **Hylocichla aliciæ Baird.** GRAY-CHEEKED THRUSH. *Alice's Thrush.* — A not uncommon migrant. There is one in the Smith collection, taken at Johnston on September 23, 1889.

---

[1] Bull. Nut. Orn. Club, Vol. 2, No. 1, p. 21.
Birds of Conn. Merriam. Trans. Conn. Acad. Vol. IV, p. 9.
Allen's Revised List Birds Mass., Bull. Am. Mus. Nat. Hist., Vol. I, p. 261.
Coues and Stearns's New Eng. Bird Life, Part I, p. 80, 81.
Allen's List Birds Mass.: with Anno. Bull. Essex Inst. Vol. X, p. 3.

(286) 757a. **Hylocichla aliciæ bicknelli (Ridgw.).** BICK-NELL'S THRUSH. — A not uncommon migrant. Mr. H. S. Hathaway took a young male on October 9, 1898, at Warwick,[1] now in the Smith collection.
(May 10 to 20), (September 28) to October 9.

(287) 758a. **Hylocichla ustulatus swainsonii (Cab.).** OLIVE-BACKED THRUSH. — A not uncommon migrant.
May 8 to (June), (September 10) to October 2.

(288) 759b. **Hylocichla aonalaschkæ pallasii (Cab.).** HERMIT THRUSH. — A common migrant. Mr. H. S. Hathaway records one in winter from Escoheag.[2]
April 19 to (May 5) (October 5) to October 20 (November 20), winter.

(289) 761. **Merula migratoria (Linn.).** AMERICAN ROBIN. *Migratory Thrush.* — An abundant summer, and winter resident.*
March 13 to November 11. Winter.

(290) 766. **Sialia sialis (Linn.).** BLUEBIRD. — A common summer resident. Stray birds, and small flocks are occasionally seen during the winter.†
March 2 to November 27. Winter.

---

[1] Osprey, Vol. III, No. 7, p. 110.
[2] Osprey, Vol. III, No. 7, p. 111.
* Ran Notes, Vol. III, No. 4, p. 27.
F. & S., Vol. XXII, No. 5, p. 183.
Nid., Vol. II, No. 12, p. 170.
† Ran. Notes, Vol. III, No. 4, p. 27.
F. & S., Vol. 6, No. 17, p. 266, and Vol. XXII. No. 9, p. 203.

EXTIRPATED SPECIES.

(1) 305. **Tympanuchus cupido** (Linn.). HEATH HEN. The assumption is that the species formerly common here was the Heath Hen, not the Prairie Hen of the West. Dr. Rives states that "The Pinnated Grouse or Prairie Hen, once very common all over this part of the country, is now no longer found east of the Ohio River, with the exception of a few which, I believe, are still in existence on the Islands of Martha's Vineyard and Naushon,"[1] and Mr. J. M. Southwick also makes mention of their occurrence.[2] Since Dr. Rives' paper appeared it has been shown that the Martha's Vineyard bird was different from the Western Prairie Hen now known as *Tympanuchus americanus*.[3] The Heath Hen is now confined to the island of Martha's Vineyard, where it is at present nearly, if not, extinct. There is a record of no value that it may be well to mention here of the "Pinnated Grouse" in the State in 1897, by Mr. Edwin R. Lewis. The bird was undoubtedly (*T. americanus*) introduced[4] from the West.

(2) 310. **Meleagris gallopavo fera Vieill.** WILD TURKEY. — Formerly common, as it was known to be, all over Southern New England.[5]

(3) 315. **Ectopistes migratorius** (Linn.). PASSENGER PIGEON. *Wild Pigeon.* — Formerly a common migrant. Mr. Newton Dexter presented in 1861 a pair to the Franklin Society collection. Col. Powel includes it in his List of Birds shot near Newport (1883-4).[6] It is stated that Mr. Walter Angell saw a flock of eight in August, 1893. The last bird, however, taken within the State seems to have been killed in 1886.

---

[1] Dr Rives' List, p. 31.
[2] Southwick's List, p. 4.
[3] Auk, Vol. II, No. 1, p. 80.
[4] F. & S., Vol. XLVIII, No. 2, p. 285.
[5] Bull. Nut. Orn. Club, Vol. 1, No. 1, p. 55.'
[*] Dr. Rives' List, p. 31.
Southwick's List, p. 5.
[6] Col. Powel's List, p. 42.

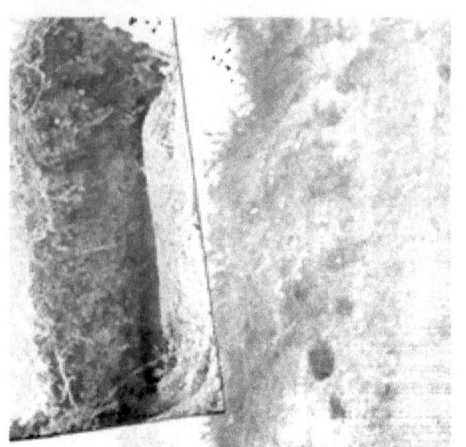

HYPOTHETICAL LIST.

Containing species for which we have some, but not conclusive, evidence of their occurrence.

(1) 13. **Fratercula arctica (Linn.).** PUFFIN. — Dr. William C. Rives in his paper says "if I am not mistaken have been found here" (Newport).[1]

(2) 42. **Larus glaucus Brünn.** GLAUCOUS GULL. *Burgomaster.* — Included in Col. Powel's list.[2]

(3) 62. **Xema sabinii (Sab.).** SABINE'S GULL. *Forked-tailed Gull.* — Included in Col. Powel's list.[3]

(4) 63. **Helochelidon nilotica (Hasselq.).** MARSH TERN. *Gull-billed Tern.* — Included in Col. Powel's list[4] as taken near Newport.

(5) 159. **Somateria mollissima borealis C. L. Berhm.** NORTHERN EIDER. *Greenland Eider.* There was found in the collection of Mr. R. L. Agassiz who made the greater part of his collection at Newport, a bird in a tray with two King Eiders (*Somateria spectabilis*) taken at Newport in December, 1885, and although without a label, probably taken with these two birds. The specimen is now in the collection of the Museum of Comparative Zoology, Cambridge, Mass.

(6) 211. **Rallus crepitans Gmel.** CLAPPER RAIL. — Included in Col. J. H. Powel's List.[6]

---

[1] Rives' List, p. 41.
[2] Col. Powel's List, p. 42.
[3] Col. Powel's List, p. 42.
[4] Col. Powel's List, p. 42.
[5] Col. Powel's List, p. 42.
[6] Coues and Stearns's, New Eng. Bird Life, Part II, p. 279.

(7) 244. **Tringa ferruginea Brünn.** CURLEW SANDPIPER. — Included in Col. Powel's List.[1]

(8) 280. **Ægialitis wilsonia (Ord.).** WILSON'S PLOVER. — Included in Col. Powel's List as a doubtful species.[1]

(9) 301. **Lagopus lagopus (Linn.).** WILLOW PTARMIGAN. *White Partridge.* — It is reported that several White Partridges, supposedly Ptarmigans, were seen during the winter of 1887.

(10) 735a **Parus atricapillus septentrionalis (Harris).** LONG-TAILED CHICKADEE. — A straggler from the West. Mr. J. M. Southwick writes that he shall soon have in the Smith collection a Long-tailed Chickadee, killed in Johnston by a brother of Mr. Walter Angell. Full data is now with the bird, which is mounted. This specimen was for some time lost and lately again found. It has been thought best not, as yet, to include this species in the Annotated List.

---

[1] Col. Powel's List, p. 42.

Rose-breasted Grosbeak's Nest.

Photographed by Mr. H. S. Hathaway, May 31, 1899, at Warwick.

BIBLIOGRAPHY.

1867 Samuels, Edward S. "Ornithology and Oology of New England." Nichols and Noyes, Boston.

1869 Allen, J. A. "Notes on Some of the Rarer Birds of Massachusetts." 1864–5 Gerfalcon record. Amer. Nat., Vol. III, No. 10, p. 513.

1869 Samuels, Edward S. "The Birds of New England and Adjacent States." Several editions. Noyes, Holmes & Co., Boston.

1870 Brewer, T. M. "Sea-side Ornithology." Amer. Nat., Vol. III, No. 4.

1871 Collete, J. R. "The Osprey." Amer. Nat., Vol. IV, No. 1, p. 57.

1873 Editors. Quail and Snipe. F. & S., Vol. I, No. 15, p. 235.

1874 Baird, Brewer and Ridgway. "History of North American Birds." 1864–5 Gerfalcon record. Vol. III, p. 115

1874 Gould, Stephen. "English Sparrow." Introduction into Newport. Amer. Nat., Vol. VIII, No. 11, p. 692.

1876 Howland, J. S. "Arrival Notes." F. & S., Vol. 6, No. 17, p. 266.

1876 "Shot." Snipe, Summer Yellow-legs and Dowitchers at Newport. F. & S., Vol. 6, No. 23, p. 376.

1875 Brewer, T. M. "Catalogue of the Birds of New England and Adjacent Localities." Pro. Boston Soc. of Nat. Hist., Vol. XVII, Mar. 3, p. 436.

1875 Editors. "Mallard." One shot at Newport Nov. 1, 1875. F. & S., Vol. V, No. 13, p. 204.

1877 Editors. "The Little Auk." F. & S., Vol. 7, No. 25, p. 388.

1877 Merriam, C. Hart. "The Birds of Connecticut." Trans. of Conn. Acad., Vol. IV, p. 9.

1877 Minot, H. D. "The Land-Birds and Game-Birds of New England." Naturalist's Agency, Salem, Mass. Estes & Lauriat, Boston.

1877 Purdie, H. A. "Notice of a Few Birds of Rare or Accidental Occurrence in New England." Bull. Nut. Orn. Club, Vol. II, No. 1, p. 22.

1878 Allen, J. A. "A List of the Birds of Massachusetts, with Annotations." Bull. Essex Institute, Vol. X, p. 3.

1878 Editors. Snipe-winter record. F. & S., Vol. 9, No. 26, p. 489.

1879 Deane, Ruthven. "Additional Capture of the Caerulean Warbler in New England." Bull. Nut. Orn. Club, Vol. IV, No. 3, p. 185.

1879 Dexter, Newton. "Capture of a Swan in Rhode Island." F. & S., Vol. XIII, No. 17, p. 848.

1879 Peckham, B. J. "Sterna caspia in Rhode Island." Ool. Vol. 5, No. 4, p. 32.

1880 Allen, J. A. "Capture of a South American Finch near Providence, R. I." Bull. Nut. Orn. Club, Vol. V, No. 4, p. 240.

1880 Deane, Ruthven. "The Large-billed Water Thrush in Eastern Rhode Island." Bull. Nut. Orn. Club, Vol. V, No. 2, p. 116.

1880 Deane, Ruthven. "The Little Blue Heron in Rhode Island." Bull. Nut. Orn. Club, Vol. V, No. 2, p. 123.

1880 F. Winter record of Woodcock and Florida Gallinule. F. & S., Vol. 15, No. 19, p. 371.

1880 Jencks, Fred. T. "Capture of the Carolina Wren and other Rare Birds in Rhode Island." Bull. Nut. Orn. Club, Vol. V, No. 4, p. 237.

1880 Jencks, Fred. T. "Capture of the Hudsonian Titmouse in Rhode Island." Bull. Nut. Orn. Club, Vol. VI, No. 1, p. 54.

1880 Jencks, Fred. T. "Least Bittern." O. & O., Vol. 5, No. 10, p. 78.

1880 Occasional. Teal and Broadbill, etc., at Newport. F. & S., Vol. 15, No. 14, p. 271.

1880 Slade, Elisha. "Notes on the Fish Hawks." Amer. Nat., Vol. XIV, No. 7, p. 528.

1880 X Y Z. Record of Wild Geese. F. & S., Vol. 15, No. 20, p. 389.

1880 X Y Z. Winter record for Snipe. F. & S., Vol. 15, No. 22, p. 43.

1881 Brewster, William. "Early arrival in New England of the Least Bittern." Bull. Nut. Orn. Club, Vol. VI, No. 3, p. 186.

1881 Bumpus, H. C. "Breeding Habits of the Fish Hawk." Am. Nat., Vol. XV, No. 10, p. 809.

1881 Bumpus, H. C. Record of Richardson's Owl. O. & O., Vol. 6, No. 2, p. 14.

1881 Collins, W. H. "Note on Least Bittern." O. & O., Vol. 6, No. 1, p. 8.

1881 Coues and Stearns. "New England Bird Life," part I. Various Notes.

1881 Editors. "Caspian Tern in Rhode Island, a correction." O. & O., Vol. 6, No. 7, p. 52.

1881 Jencks, Fred. T. "Catching a Tartar." O. & O., Vol. 6, No. 2, p. 14.

1881 Jencks, Fred. T. "Least Bittern." O. & O., Vol. 6, No. 1, p. 6.

1881 Jencks, Fred. T. "Richardson's Owl in Rhode Island." Bull. Nut. Orn. Club, Vol. VI, No. 2, p. 123.

1881 Skinner, Fred'k. "Canvas-Backs in Rhode Island." F. & S., Vol. 15, No. 23, p. 447.

1881 Southwick and Jencks. "Addition to the Rhode Island fauna." O. & O., Vol. 6, No. 6, p. 44.

1882 Jencks, Fred. T. "Large-billed Water Thrush." O. & O., Vol. 7, No. 15, p. 114.

1882 Jencks, Fred T. "Purple Gallinule in Rhode Island." Bull. Nut. Orn. Club, Vol. VII, No. 2, p. 124.

1882 X. Brant in Narragansett Bay. F. & S., Vol. XVIII, part II. No. 6, p. 107.

1883 Jencks, Fred. T. "Capture of the Richardson's Owl near Providence, R. I." Bull. Nut. Orn. Club, Vol. VIII, No. 2, p. 122.

1883 Jencks, Fred. T. "Duck Hawks." O. & O., Vol. 8, No. 12, p. 92.

1883 Jencks, Fred. T. "Great Gray Owl in Rhode Island." Bull. Nut. Orn. Club, Vol. VIII, No. 3, p. 183.

1883 Jencks, Fred. T. Gerfalcon killed at Pt. Judith 1883. O. & O., Vol. 8, No. 12, p. 91.

1883 Jencks, Fred. T. "The Baldpate in Rhode Island." Bull. Nut. Orn. Club, Vol. VIII, No. 1, p. 62.

1883 Southwick and Jencks. Capture of Sparrow Hawk and Snowy Owls. O. & O., Vol. 8, No. 3, p. 24.

1883-4 Powel, Col. J. H. "List of Birds Shot near Newport." Pro. New. Nat. His. Soc. 1883-4, p. 42.

1883-4 Rives, William C., M. D. "The Birds of Newport." Pro. New. Nat. His. Soc. 1883-4, p. 28. Also in Newport Daily News. March 11, 1884.

1884 Aldrich, T. M. Winter Notes. F. & S., Vol. XXII, No. 5, p. 83.

1884 Dexter, S. F. Albinos. F. & S., Vol. XXII, No. 9, p. 165.

1884 Editors. "A New Bird for Rhode Island and Second for New England." Ran. Notes, Vol. I, No. V, p. 8.

1884 Editors. "A Very Rare Bird in Rhode Island." Ran. Notes, Vol. I, No. I, p. 6.

1884 Editors. "A Nonpareil in Rhode Island." Ran. Notes, Vol. I, No. V, p. 8.

1884 Editors. "An English Corn-Crake in Rhode Island." Ran. Notes, Vol. I, No. VI, p. 3.

1884 Editors. "Common Cormorant in Rhode Island." Ran. Notes, Vol. II, No. V, p. 34.

1884 Editors. "Curious Death of a Saw-whet Owl." Ran. Notes, Vol. I, No. I, p. 4.

1884 Editors. "Early Woodcock." Ran. Notes, Vol. I, No. V, p. 8.

1884 Editors. "Great Gray Owl, Spectral Owl." Ran. Notes, Vol. I, No. VII, p. 3.

1884 Editors. "Late Crossbills, L'americana." (*Sic.*) Ran. Notes, Vol. I, No. VI, p. 6.

1884 Editors. "Night Heron in Winter." Ran. Notes, Vol. I, No. I, p. 9.

1884 Editors. Note on Connecticut Warbler, Scarlet Tanager, and Maryland Yellow-throat. Ran. Notes, Vol. I, No. XI, p. 6.

1884 Editors. "Prothonotary Warbler." Ran. Notes, Vol. I, No. VI, p. 3.

1884 Editors. Two Brünnichs Murres and Night Heron, winter record. Ran. Notes, Vol. I, No. II, p. 8.

1884 F. H. "Rhode Island Game." F. & S., Vol. XXI, No. 25, p. 498.

1884 Hazard, R. G. 2nd. "The Occurrence of the Golden Swamp Warbler in Rhode Island." Auk, Vol. I, No. 3, p. 290.

1884 Jencks, Fred. T. "Another Gyrfalcon in Rhode Island." Auk, Vol. I, No. I, p. 94.

1884 Jencks, Fred. T. "A Trip to a Heronry." O. & O., Vol. 9, No. 8, p. 103.

1884 Jencks, Fred. T. "Birds During January" and "February." Prov. Journal, Feb. 7.

1884 Jencks, Fred. S. "Birds during March" and "April" Prov. Journal, April 7.

1884 Talbot, H. A. "A Trip to a Heronry." O. & O., Vol. 9, No. 7, p. 80.

1884 Talbot, H. A. "Brünnich's Guillemot in Rhode Island." Ran. Notes, Vol. I, No. I, p. 6.

1884 Talbot, H. A. "Notes from Warwick Neck, R. I." O. & O., Vol. 9, No. 5, p. 58.

1884 Water Fowl. Wintering of Tree Swallows, Bluebirds, etc. at So. Kingston. F. & S., Vol. XXII, No. 9, p. 203.

1885 Brown, F. C. "Another Richardson's Owl in Mass." Reference to Jencks' birds of 1881-82. Auk, Vol. II, No. 4, p. 384.

1885 C. H. L. "Spring Notes." F. & S., Vol. XXIV, No. 12, p. 225.

1885 Editors. "A King Eider." Ran. Notes, Vol. I, No. II, p. 9.

1885 Editors. Snowy Owl at Newport. Ran. Notes, Vol. II, No. III, p. 23.

1885 J. A. A. "Rives on the Birds of Newport, R. I." Auk, Vol. II, No. 2, p. 208.

1885 Southwick and Jencks. "Snowy Owl." O. & O., Vol. 10, No. 3, p. 48.

1885 Editors. "The Ipswich Sparrow in Rhode Island." Ran. Notes, Vol. II, No. III, p. 17.

1885 W. M. H. "Fowl in Rhode Island." F. & S., Vol. XXIV, No. 12, p. 228.

1886 Allen, J. A. "A Revised List of the Birds of Massachusetts." Bull. Amer. Museum of Nat. History. New York. Vol. I, No. 7, p. 221.

1886 American Ornithologists' Union. "Check-List of North American Birds." New York.

1886 Editors. Capture of Barn Owl and Razor-billed Auk, Ran. Notes, Vol. III, No. XII, p. 91.

1886 Editors. "Arrival Notes." Ran. Notes, Vol. III, No. IV. p. 27.

1886 Editors. Capture of Yellow-crowned Night Heron at Tiverton. Ran. Notes, Vol. III, No. VII, p. 49.

1886 Editors. Note on a scarcity of Snowy Owls. Ran. Notes, Vol. III, No. II, p. 9.

1886 Brewster, Wm. "Occurrence of the Prothonotary Warbler in Rhode Island." Auk, Vol. III, No. 3, p. 411.

1886 Editors. "Purple Gallinule." Ran. Notes, Vol. III, No. X, p. 79.

1886 Editors. "Seasonable Notes." Ran. Notes, Vol. III, No. V, p. 37.

1886 Editors. "Tennessee Warbler in Rhode Island." Ran. Notes, Vol. III, No. X, p. 79.

1886 Editors. Wintering of Brown Thrasher in Rhode Island. O. & O., Vol. 11, No. 6, p. 84.

1886 F. B. W. Black Guillemot. O. & O., Vol. 11, No. 1, p. 16.

1886 T. M. A. Grouse, Quail and Woodcock Note. F. & S., Vol. XXVI, No. 25, p. 489.

1887 Andros, Fred. W. "A List of Birds of Bristol County, Mass." O. & O., Vol. 12, No. 7, p. 138. Note.

1887 Editors. Capture of Yellow Rail. O. & O., Vol. 12, No. 2, p. 32.

1887 Baird, S. F. "Occurrence of Cory's Shearwater and several species of Jaegers in large flocks in the vicinity of Gay Head, Mass., during the autumn of 1886." Auk, Vol. IV, No. 1, p. 71.

1887 Dexter, Newton. "Golden Eagle in Rhode Island." F. & S., Vol. XXVIII, No. 6, p. 106.

1887 Dwight, Dr. Jonathan, Jr. "A New Race of Sharp-tailed Sparrows." Auk, Vol. IV, No. 2, p. 136.

1887 G. B. R. Two Killdeers taken at Newport. F. & S., Vol. XXVIII, No. 12, p. 249.

1887 Rives, Wm. C., Jr., M. D. "Wilson's Phalarope in Rhode Island." Auk, Vol. IV, No. 1, p. 73.

1888 Dexter, Newton. "Whose Hawk was This?" F. & S., Vol. XXXI, No. 15, p. 285.

1888 Lawton, Chas. H. "The Water-Birds of Newport." Pro. New. Nat. His. Soc., 1888, p. 16.

1888 Rives, Wm. C., Jr., M. D. "Cory's Shearwater at Newport, R. I." Auk, Vol. V, No. 1, p. 108.

1888 Southwick, James M. "Our Birds of Rhode Island." Pro. New. Nat. His. Soc., 1888, p. 3.

1888 Trumbull, Gurdon. "Names and Portraits of Birds." Harper & Bros., New York, 1888. R. I. local names.

1888 White, Harry Gordon. Mocking-Bird at Newport. O. & O., Vol. 13, No. 12, p. 192.

1889 Chadbourne, Dr. A. P. "An unusual Flight of Killdeer Plover (*Ægialitis vocifera*) along the New England Coast." Auk, Vol. VI, No. 3, p. 255.

1889 Dexter, Newton. "Rare Birds in Rhode Island." F. & S., Vol. XXXIII, No. 19, p. 364.

1889 Jencks. Fred. T. "A Little Brown Crane in Rhode Island." Independent Citizen, Nov. 16. Providence.

1889 Robinson, Lieut. Wirt. "Some Rare Rhode Island Birds." Auk, Vol. VI, No. 2, p. 194.

1889 Southwick, J. M. "Capture of the American Egret on Prudence Island." O. & O., Vol. 14, No. 4, p. 63.

1889 Southwick, J. M. Capture of the Little Brown Crane. O. & O., Vol. 14, No. 10, p. 159.

1890 Brewster, Wm. "The Little Brown Crane in Rhode Island." Auk, Vol. VII, No. 1, p. 89.

1890 E. Woodcock's Notes. F. & S., Vol. XXXV, No. 16, p. 312.

1890 Mackay, G. H. "Somateria dresseri. — The American Eider." Auk, Vol. VII, No. 4, p. 318.

1890 Southwick and Critchley. "Red Phalarope in Rhode Island." O. & O., Vol. 15, No. 11, p. 166.

1890 F. L. G. "Wilson's Snipe in Rhode Island." F. & S., Vol. XXXVI, No. 6, p. 105.

1891 Jencks, Fred. T. "A Rare Capture in Rhode Island." The Collector's Monthly, Vol. 2, No. 12. p. 72.

1891 Mackay, G. H. "The Scoters in New England." Auk, Vol. VIII, No. 3, p. 279.

1891 Miller, G. S., Jr. "Further Cape Cod Notes." Auk, Vol. VIII, No. 1, p. 117.

1891 Southwick and Critchley. "Early Arrival of Scarlet Tanager at Apponaug, R. I." O. & O., 16, No. 5, p. 78.

1892 Dunn, Claude. "The Caspian Tern at Rhode Island." O. & O., Vol. 17, No. 6, p. 76.

1892 Mackay, G. H. 'Gull Dick.' Auk, Vol. IX, No. 2, p. 227.

1892 Southwick, J. M. "Rhode Island Birds." Prov. Journal. Feb. 28.

1892 Taylor, A. O. D. "Occurrence of the Black Gyrfalcon in Rhode Island." Auk, Vol. IX, No. 4, p. 300.

1893 Dunn, C. G. "An American Egret taken in Rhode Island." O. & O., Vol. 18, No. 6, p. 94.

1893 Glezen, F. L. Capture of Blue Goose at Charlestown Beach. Providence Journal, Jan. 25.

1893 Glezen, F. L. "Blue Goose in Rhode Island." F. & S. Vol. XL, No. 3, p. 48.

1893 Mackay, G. H. "Larus argentatus smithsonianus." Auk, Vol. X, No. 1, p. 76.

1894 Editors. "A Quail's Nest in October." Prov. Journal, Oct. 27.

1894 L. A. C. "Transplanting Quail." F. & S., Vol. XLII, No. 12, p. 248.

1894 Livermore, J. "The Yellow-crowned Night Heron in Rhode Island." Auk, Vol. XI, No. 2, p. 177.

1894 Mackay, G. H. "Further News of the 'Gull Dick.'" Auk, Vol. XI, No. 1, p. 73.

1894 Mackay, G. H. "Habits of the Double-crested Cormorant in Rhode Island." Auk, Vol. XI, No. 1, p. 18.

1894 Roberts, S. H. "Rhode Island Quail." F. & S., Vol. XLII, No. 14, p. 291.

1894 Schuyler, E. O. "A Quail's Nest in October." F. & S., Vol. XLIII, No. 17, p. 355.

1894 Schuyler, E. O. "Quail's Nest with ten eggs in October." Prov. Journal, October 28.

1894 Tode. Restocking Rhode Island with Quail. F. & S., Vol. XLII, No. 3, p. 49.

1895 American Ornithologists' Union. "Check-List of North American Birds." Second and Revised Edition. New York.

1895 Brewster, Wm. "A Remarkable Flight of Pine Grosbeaks (Pinicola enucleator)." Auk, Vol. XII, No. 3, p. 245.

1895 Brewster, William. Minot's "The Land-Birds and Game-Birds of New England." Revised second edition. Houghton, Mifflin & Co., Boston. Footnotes on distribution.

1895 Chapman, Frank M. "Handbook of Birds of Eastern North America." Several editions. D. Appleton & Co., New York.

1895 Doe, Chas. E. Killdeer Nesting. Nid. Vol. II, No. 12, p. 179.

1895 Howe, R. H., Jr. "A Large Brood of Ospreys." Auk, Vol. XII, No. 4, p. 389.

1895 Howe, R. H. Jr. "Ospreys at Bristol, R. I." Auk, Vol. XII, No. 3, p. 300.

1895 Mackay, G. H. "'Gull Dick' Again." Auk, Vol, XII, No. 1, p. 76.

1895 Newbury, F. E. "Three Families — One Hole." Nid. Vol. II, No. 12, p. 170.

1896 Brewster, Wm. "Occurrence of Wood Ibis in Bristol County, Mass." Auk, Vol. XIII, No. 3, p. 341.

1896 G. C. Note on Bluebirds. F. & S., Vol. XLVI, No. 15, p. 293.

1896 Hathaway, H. S. "A Wood Ibis in Rhode Island." Osprey, Vol. I, No. 4, p. 67.

1896 Howe, R. H., Jr. "Four Winter Records of the Short eared Owl on the Mass. Coast." Auk, Vol. XIII, No. 3, p. 257.

1896 Howe, R. H., Jr. "A List of the Birds of Bristol, R. I. and Adjacent Localities." Bristol Phoenix, April 10.

1896 Howe, R. H., Jr. "Every Bird." Bradlee Whidden, Boston. Various Notes.

1896 Mackay, G. H. "'Gull Dick' Again." Auk, Vol. XIII, No. 1, p. 78.

1896 Nuthatch. "A Nuthatch's Device." F. & S., Vol. XLVI, No. 6, p. 215.

1896 Nuthatch. "Rhode Island Bird Notes." F. & S., Vol. XLVI, No. 10, p. 195.

1896–7 Newbury, F. E. "Finding the Killdeer's Nest." Nid., Vol. IV, No. 3, 4, 5, p. 43.

1897 Howe, R. H., Jr. "The Sea-side Sparrow at Middletown, R. I." Auk, Vol. XIV, No. 2, p. 219.

1897 Howe, R. H., Jr. "The Terns of Dyer's and the Weepecket Islands." Auk, Vol. XIV, No. 2, p. 203.

1897 Hughes, W. M. "Wintering of Robins, etc. at Portsmouth." F. & S., Vol. XLVIII, No. 11, p. 204.

1897 Lewis, Edwin R. "Pinnated Grouse in Rhode Island." F. & S. Vol. XLVIII, No. 20, p. 285.

1897 Sturtevant, Edward. "The Sea-side Sparrow at Middletown, R. I." Auk, Vol. XIV, No. 4, p. 322.

1897 W. H. M. Rhode Island Game. F. & S., Vol. XLVIII, No. 4, p. 90.

1897 W. H. M. Woodcock Note. F. & S., Vol. XLVIII, No. 13, p. 249.

1897 W. H. M. The Taking of Widgeon, F. & S., Vol. XLIX, No. 18, p. 348.

1898 Bent, A. C. "Black Gyrfalcon in Rhode Island." Auk, Vol. XV, No. 1, p. 54.

1898 Hathaway, H. S. "A Rare Bird in Rhode Island." Osprey, Vol. 2, No. 6, 7, p. 91.

1898 Mackay, G. H. 'Gull Dick.' Auk, Vol. XV, No. 1, p. 49.

1898 Wright, N. M. "Rhode Island Birds." Prov. Journal, Sept. 25.

1899 Hathaway, H. S. "Birds Wintering in Rhode Island." Osprey, Vol. III, No. 7, p. 111.

1899 Hathaway, H. S. "Rare Birds in Rhode Island." Osprey, Vol. III, No. 7, p. 110.

1899 Hazard, R. G. "The Carolina Wren at Peacedale, R. I." Auk, Vol. XVI, No. 1, p. 83.

1899 Howe, R. H., Jr. "On the Birds' Highway." Small, Maynard & Co., Boston. Containing two chapters and an annotated list of 82 species observed at Bristol, R. I. pp. 12, 69, 153.

1899 Howe, R. H., Jr. "Notes from Rhode Island." Auk, Vol. XVI, No. 2, p. 189.

1899 Howe, R. H., Jr. "Sexual Difference in Size of the Pectoral Sandpiper." Auk, Vol. XVI, No. 2, p. 179.

1899 Howe, R. H., Jr. "Revival of the Sexual Passion in Birds in Autumn." Auk, Vol. XVI, No. 3, p. 286.

1899 Sturtevant, Edward. "The Carolina Wren Breeding in Rhode Island." Auk, Vol. XVI, No. 3, p. 284.

1899 Robinson, Wirt. "*Ammodramus henslowii.* A correction." Auk, Vol. XVI, No. 4, p. 356.

ERRATA, ADDITIONS, ETC.

Page 11. "(*Americana deglandi* and *perspicillata*)" should read, (*americana*, *deglandi* and *perspicillata*).
Page 13. "Woodcock" should read, American Woodcock.
Page 14. "Flicker" should read, Northern Flicker.
Page 15. "*Helminthophila ruficapilla*" should read, *Helminthophila rubricapilla*.
Page 17. "and the land around the rock" should read, and the band around the rock.
Page 21. "Stolid Sandpiper" should read, stolid Sandpiper.
Page 30. The following reference should also be added to those for the 1876 Sooty Tern (*Sterna fuliginosa*) record, Allen's List Birds Mass. with Anno. Bull. Essex Inst. Vol. X, p. 30.
Page 34. "Col. J. H. Powell" should read, Col. J. H. Powel.
Page 35. "*Anas strepera* Linn." should read, *Chaulelasmus streperus* (Linn.).
Page 45. Little Brown Crane. The specimen has just been placed in the Smith collection.
Page 49. "W. Hare H. Powel" should read, H. W. Hare Powel.
Page 85. "² Bull. Nut. Orn. Club, Vol. VI, No. — p. 54." should read, ² Bull. Nut. Orn. Club, Vol. VI, No. 1, p. 54.

(291) **597 Guiraca cærulea** (Linn.). BLUE GROSBEAK.— We have just received word from Mr. F. T. Jencks of the capture of a young bird by him on his farm in Drownville, on October 12, 1899. The bird was in some blackberry bushes, and from its actions attracted his attention. Mr. Jencks sent the bird to Mr. J. M. Southwick. It will undoubtedly be placed in the Smith collection. This is the first capture of this species in the State.

INDEX.

SCIENTIFIC NAMES.

Acanthis linaria, 70.
    linaria rostrata, 70.
Accipiter atricapillus, 57.
    cooperi, 13, 57.
    velox, 13, 57.
Actitis macularia, 13, 21, 22, 53.
Ægialitis meloda, 13, 55.
    semipalmata, 55.
    vocifera, 13, 54.
    wilsonia, 90.
Agelaius phoeniceus, 14, 67.
Astragalinus tristis, 14, 70.
Aix sponsa, 13, 37.
Anorthura, hiemalis, 84.
Alca torda, 27.
Alle alle, 27.
Ammodramus caudacutus, 15, 72.
    caudacutus subvirgatus, 72.
    henslowii, 71.
    maritimus, 15, 72.
    nelsoni, 72.
    princeps, 71.
    sandwichensis savanna, 14, 71.
    savannarum passerinus, 14, 71.
Ampelis cedrorum, 15, 76.
Anas boschas, 22, 35.
    obscura, 13, 22, 35.
Anthus pensilvanicus, 83.
Antrostomus vociferus, 14, 65.
Aquila chrysaëtos, 58.
Archibuteo lagopus sancti-johannis, 58.
Ardea candidissima, 44.
    cærulea, 44.
    egretta, 43.
    herodias, 42.
    virescens, 13, 44.
Ardetta, exilis, 13, 43.
Arenaria interpres, 20, 22, 55.
Asio accipitrinus, 14, 61.
    wilsonianus, 14, 61.
Aythya affinis, 38.
    americana, 37.
    collaris, 38.
    marila, 38.
    vallisneria, 37.

BARTRAMIA longicauda, 53.
Bonasa umbellus, 13, 56.
Botaurus lentiginosus, 13, 43.
Branta bernicla, 41.
    canadensis, 41.
Bubo virginianus, 14, 63.
Buteo borealis, 13, 57.
    latissimus, 13, 58.
    lineatus, 13, 57.

CALCARIUS lapponicus, 71.
Calidris arenaria, 51.
Carpodacus purpureus, 14, 69.
Cathartes aura, 56.
Cepphus, grylle, 27.
Certhia familiaris fusca, 85.
Ceryle alcyon, 14, 64.
Chætura pelagica, 14, 65.
Charadrius dominicus, 54.
Charitonetta albeola, 39.
Chaulelasmus streperus, 35.
Chen cærulescens, 41.
    hyperborea, 41.
    hyperborea nivalis, 41.
Chordeiles virginianus, 14, 65.
Circus hudsonius, 13, 56.
Cistothorus palustris, 16, 85.
    stellaris, 16, 85.
Clangula clangula americana, 38.
Clivicola riparia, 15, 76.
Coccyzus americanus, 14, 63.
    erythrophthalmus, 14, 64.
Colaptes auratus luteus, 14, 65.
Colinus virginianus, 13, 55.
Colymbus auritus, 21, 25.
    holbœllii, 21, 25.
Compsothlypis americana usneæ, 15, 79.
Contopus borealis, 66.
    virens, 14, 66.
Corvus americanus, 12, 14, 67.
Crex crex, 46.
Crymophilus fulicarius, 47.
Cyanocitta cristata, 14, 67.
Cyanospiza ciris, 74.
    cyanea, 15, 74.

DAFILA, acuta, 35, 37.
Dendroica æstiva, 15, 79.
    blackburniæ, 80.
    cærulescens, 79.
    castanea, 80.
    coronata, 79.
    discolor, 15, 81.
    maculosa, 79.
    palmarum, 80.
    palmarum hypochrysea, 81.
    pensylvanica, 15, 80.
    rara, 80.
    striata, 80.
    tigrina, 79.
    vigorsii, 15, 80.
    virens, 15, 80.
Dolichonyx oryzivorus, 14, 67.
Dryobates pubescens medianus, 14, 64.
    villosus, 14, 64.

ECTOPISTES migratorius, 88.
Empidonax flaviventris, 66.
    minimus, 14, 66.
Ereunetes occidentalis, 51.
    pusillus, 50, 51.
Erismatura jamaicensis, 13, 41.

FALCO columbarius, 60.
    gyrfalco islandicus, 59.
    peregrinus anatum, 59.
    rusticolus gyrfalco, 59.
    rusticolus obsoletus, 59.
    sacer, 59.
    sparverius, 13, 60.
Fratercula arctica, 89.
Fulica americana, 47.

GALEOSCOPTES carolinensis, 15, 83.
Gallinago delicata, 13, 48.
Gallinula galeata, 13, 47.
Gavia imber, 21, 26.
    lumme, 21, 25, 26.
Geothlypis agilis, 81.
    formosa, 81.
    philadelphia, 82.
    trichas, 15, 82.
Grus canadensis, 45.
Gubernatrix cristatella, 75.

HALIÆETUS leucocephalus, 58.
Harelda hyemalis, 22, 39.
Harporhyncus rufus, 15, 83.
Helminthophila celata, 78.
    chrysoptera, 78.
    peregrina, 78.
    pinus, 15, 78.

Helminthophila rubricapillus, 15, 78.
Helochelidon nilotica, 89.
Helodromus solitarius, 52.
Hirundo erythrogastra, 15, 75.
Histrionicus histrionicus, 39.
Hydrochelidon nigra surinamensis, 31.
Hylocichla mustelinus, 16, 86.
    fuscescens, 16, 86.
    fuscescens salicicola, 86.
    aliciæ, 86.
    aliciæ bicknelli, 87.
    ustulatus swainsonii, 87.
    aonalaschkæ pallasii, 87.

ICTERIA virens, 15, 82.
Icterus galbula, 14, 68.
    spurius, 14, 68.
Ionornis martinica, 46.

JUNCO hyemalis, 73.

LAGOPUS lagopus, 90.
Lanius borealis, 76.
    ludovicianus excubitorides, 76.
Larus argentatus smithsonianus, 22, 28.
    atricilla, 29.
    glaucus, 89.
    marinus, 22, 28.
    philadelphia, 22, 28.
Limosa fedoa, 51.
    haemastica, 51.
Lophodytes cucullatus, 35.
Loxia curvirostra minor, 70.
    leucoptera, 70.

MACRORHAMPUS criseus, 49.
    scolopaceus, 49.
Mareca americana, 35, 36.
Megascops asio, 14, 63.
Melanerpes carolinus, 64.
    erythrocephalus, 14, 64.
Meleagris gallapavo fera, 88.
Melospiza fusciata, 15, 73.
    georgiana, 15, 73.
    lincolni, 73.
Merganser americanus, 33.
    serrator, 22, 33.
Merula migratoria, 16, 87.
Micropalama himantopus, 49.
Mimus polyglottos, 83.
Mniotilta varia, 15, 77.
Molothrus ater, 14, 67.
Myiarchus crinitus, 14, 66,

NETTION carolinensis, 36.
Numenius borealis, 54.
    hudsonicus, 54.
    longirostris, 53.
Nyctala acadica, 63.
    tengmalmi richardsoni, 62.
Nyctea nyctea, 63.
Nycticorax nycticorax nævius, 13, 44.
    violacens, 45.

OCEANODROMA leucorhoa, 32.
    oceanicus, 32.
Oidemia americana, 11, 22, 40.
    deglandi, 11, 22, 40.
    perspicillata, 11, 22, 40.
Olor columbianus, 42.
Otocoris alpestris, 67
    alpestris praticola, 67.
Pandion haliætus carolinensis, 14, 60.
Parus atricapillus, 16, 85.
    atricapillus septentrionalis, 90.
    hudsonicus, 85.
Passer domesticus, 69.
Passerella iliaca, 74.
Passerina nivalis, 70.
Petrochelidon lunifrons, 15, 75, 76.
Phalacrocorax carbo, 12, 19, 22, 33.
    dilophus, 12, 18, 19, 22, 33.
Phalaropus lobatus, 47.
Philohela minor, 13, 48.
Pinicola enucleator canadensis, 69.
Pipilo erythrophthalmus, 15, 74.
Piranga erythromelas, 15, 75.
    rubra, 75.
Podilymbus podiceps, 13, 25.
Polioptila cærulea, 86.
Pocetes gramineus, 14, 22, 71.
Porzana carolina, 13, 45.
    noveboracensis, 46.
Progne subis, 15, 75.
Pronotaria citrea, 71.
Puffinus borealis, 31, 32.
    gravis, 32.
    fuliginosus, 32.

QUISCALUS quiscula, 14, 68.
    quiscula æneus, 14, 68, 69.
Querquedula discors, 13, 36.

RALLUS crepitans, 89.
    elegans, 45.
    virginianus, 13, 45.
Regulus calendula, 86.
    satrapa, 86.

Rissa tridactyla, 28.

SARORNIS phœbe, 14, 66.
Scolecophagus carolinus, 68.
Scotiaptex cinereum, 62.
Seiurus aurocapillus, 15, 81.
    motacilla, 15, 81.
    noveboracensis, 81.
Setophaga ruticilla, 15, 83.
Sialia sialis, 16, 87.
Sitta canadensis, 85.
    carolinensis, 16, 85.
Somateria dresseri, 22, 39.
    mollissima borealis, 89.
    spectabilis, 39, 89.
Spatula clypeata, 36.
Sphyrapicus varius, 64.
Spinus pinus, 70.
Spiza americana, 74.
Spizella monticola, 73.
    pusilla, 15, 73.
    socialis, 15, 73.
Squatarola squatarola, 54.
Steganopus tricolor, 48.
Stercorarius pomarinus, 28.
Sterna antillarum, 30.
    caspia, 29.
    dougalli, 22, 30.
    forsteri, 29.
    fuliginosa, 30.
    hirundo, 13, 20, 22, 30.
Strix pratincola, 60.
Sturnella magna, 14, 68.
Sula bassana, 33.
Surnia ulula caparoch, 63.
Symphemia semipalmata, 52.
Syrnium nebulosum, 14, 61.

TACHYCINETA bicolor, 15, 76.
Tantalus loculator, 42.
Thryothorus ludovicianus, 15, 84.
Totanus flavipes, 52.
    melanoleucus, 52.
Tringa alpina pacifica, 51.
    bairdii, 50.
    canutus, 49.
    ferruginea, 90.
    fuscicollis, 50.
    maculata, 50.
    maritima, 21, 22, 50.
    minutilla, 50, 51.
Trochilus colubris, 14, 65.
Troglodytes aëdon, 16, 84.
Tryngites subruficollis, 53.
Tympanuchus americanus, 88.
    cupido, 88.
Tyrannus tyrannus, 14, 66.

URIA lomvia, 27.
    troile, 27.

VIREO flavifrons, 15, 77.
    gilvus, 15, 77.
    noveboracensis, 15, 77.
    olivaceus, 15, 77.
    philadelphicus, 77.
    solitarius, 15, 77.

WILSONIA canadensis, 82.
    mitrata, 82.
    pusilla, 82.

XEMA sabinii, 89.

ZAMELODIA ludoviciana, 15, 74.
Zenaidura macroura, 13, 56.
Zonotrichia albicollis, 72.
    leucophrys, 72.

VERNACULAR AND LOCAL NAMES.

Auk, Razor-billed, 27.

Baldpate, 36.
Beetlehead, 54.
Bittern, American, 13, 43.
    Least, 13, 43.
Blackbird, Cow, 67.
    Crow, 68.
    Red and Buff Shouldered, 67.
    Red-winged, 14, 67.
Black-breast, 54.
Blue-bill, 38.
Bluebird, 16, 87.
Bobolink, 14, 67.
Bob-White, 13, 55.
Booby, 41.
Brant, 41.
Brant-bird, 55.
Broad-bill, 38, 41.
    Bastard, 38.
    Creek, 38.
Brownback, 49.
Bunting, Black-throated, 74.
    Cow, 67.
    Indigo, 15, 74.
    Painted, 74.
    Snow, 70.
Buffle-head, 39.
Bull Bat, 65.
Butcherbird, 76.
Butter-ball, 39.
Butter-bill, 40.

Calico-bird, 55.
Canary, Wild, 79.
Canvas-back, 37.
Catbird, 15, 83.
Cedarbird, 76.
Chat, Yellow-breasted, 15, 82.
Chewink, 74.
Chickadee, 16, 85.
    Hudsonian, 85.
    Long-tailed, 90.
Chippy, 73.
Coot, 47.
    American, 47.
    Butter-bill, 40.
    Gray, 40.
    Patch-bill, 40.
    Patch-poll, 40.
    Yellow-billed, 40.
Cormorant, 30, 33.
    Common, 12, 19, 22, 33.

Cormorant, Double-crested, 12, 18, 19, 22, 34.
Cowbird, 14, 67.
Crake, Corn, 46.
Crane, 43.
    Little Brown, 45.
Creaker, 50.
Creeper, Black and White, 77.
    Brown, 85.
Crossbill, American, 70.
    Red, 70.
    White-winged, 70.
Crow, 67.
    American, 12, 14, 67.
Cuckoo, Black-billed, 14, 64.
    Yellow-billed, 14, 63.
Curlew, Eskimo, 54.
    Esquimaux, 54.
    Hudsonian, 54.
    Jack, 54.
    Long-billed, 53.

Dabchick, 25.
"Deutscher," 49.
Dickcissel, 74.
Dipper, 39.
Diver, Little, 25.
Doughbird, 54.
Dove, Carolina, 56.
    Mourning, 13, 56.
    Long-tailed, 56.
Dowitcher, 49.
    Long-billed, 49.
Dove, Sea, 27.
Dovekie, 27.
Duck, American Scaup, 38.
    American Scoter, 40.
    Black, 13, 21, 22, 35.
    Buffle-head, 39.
    Canvas-backed, 37.
    Dusky, 35.
    Eider, 39.
    Gray, 37.
    Harlequin, 39.
    Lesser Scaup, 38.
    Little Black-head, 38.
    Long-tailed, 39.
    Red-headed, 37.
    Ring-necked, 38.
    Ruddy, 13, 41.
    Shoveller, 36.
    Summer, 37.
    Surf, 40.

Duck, Velvet, 40.
    Wood, 13, 37.
EAGLE, Bald, 58.
    Golden, 58.
Egret, American, 43.
Eider, American, 22, 39.
    Greenland, 89.
    King, 39.
    Northern, 89.

FINCH, Grass, 71.
    Lincoln's, 73.
    Purple, 14, 69.
    South American, 75.
Firebird, 68.
Flicker, 14, 65.
    Northern, 65.
Flycatcher, Crested, 14, 66.
    Great-crested, 66.
    Least, 14, 66.
    Olive-sided, 66.
    Yellow-billed, 66.
"Fly-up-the-creek," 44.
Frost-bird, 54.

GADWALL, 35.
Gallinule, Purple, 46.
    Florida, 13, 47.
Gannet, 33.
Gnatcatcher, Blue-gray, 86.
Godwit, Hudsonian, 51.
    Marbled, 51.
Golden-eye, 38.
    American, 38.
Goldfinch, American, 14, 70.
Goosander, 34.
Goose, 41.
    Brant, 41.
    Blue, 41.
    Canada, 41.
    Mexican, 41.
    Lesser Snow, 41.
    Snow, 41.
    Solan, 33.
    Wild, 41.
Grackle, Bronzed, 14, 69.
    Purple, 14, 68.
    Rusty, 68.
Great May White-wings, 11, 40.
Grebe, American Red-necked, 25.
    Holboell's, 21, 25.
    Horned, 21, 25.
    Pied-billed, 13, 25.
    Red-necked, 25.
Green-head, 35, 54.
Grosbeak, Canadian Pine, 69.

Grosbeak, Pine, 69.
    Rose-breasted, 15, 74.
Grouse, Ruffed, 13, 56.
Gull, American Herring, 22, 28.
    Bonaparte's, 22, 29.
    'Dick,' 29.
    Glaucous, 89.
    Great Black-backed, 22, 28.
    Grew, 31.
    Laughing, 29.
    Mackerel, 30.
    Sabine's, 89.
    Sea, 28.
    Winter, 28.
Guillemot, Black, 27.
Gyrfalcon, Black, 59.
Gyrfalcon, 59.

HAGDON, 32.
    Black, 32.
Harlequin, 39.
Harry Wicket, 65.
Hawk, American Sparrow, 13, 60.
    American Goshawk, 57.
    American Rough-legged, 58.
    Broad-winged, 13, 58.
    Cooper's, 13, 57.
    Duck, 59.
    Fish, 60.
    Hen, 57.
    Marsh, 13, 56.
    Pigeon, 60.
    Red-shouldered, 13, 57.
    Red-tailed, 13, 57.
    Sharp-skinned, 13, 57.
Hell-diver, 25.
Hen, Heath, 88.
    Marsh, 47.
    Mud, 47.
Heron, Black-crowned Night, 13, 44.
    Great Blue, 42, 43.
    Green, 13, 44.
    Little Blue, 44.
    Night, 44.
    Snowy, 44.
    Yellow-crowned Night, 45.
High Hole, 65.
Hummingbird, Ruby-throated, 14, 65.

IBIS, Wood, 42.
Indigobird, 74.

JAEGER, Pomarine, 28.
Jay, Blue, 14, 67.
    Yellow, 65.

INDEX.

Junco, 73.
    Slate-colored, 73.

KILLDEER, 13, 54.
Kingbird, 14, 66.
Kittiwake, 28.
Kingfisher, 64.
    Belted, 14, 64.
Kinglet, Golden-crowned, 86.
    Ruby-crowned, 86.
Knot, 49.
Kreiker, 50.

LARK, Horned, 67.
    Prairie Horned, 67.
Lazy-bird, 67.
Linnet, Gray, 70.
    Red, 69.
Longspur, Lapland, 71.
Loon, 21.
    Big, 26.
    Little, 26.
    Red-throated, 21, 25.
    Tinker, 25.

MALLARD, 22, 35.
    Wild, 35.
Marlin, Common, 51.
    Ring-tailed, 51.
Martin, Purple, 15, 75.
May, White-wings, 11, 40.
Meadowlark, 14, 68.
Merganser, American, 34.
    Buff-breasted, 34.
    Hooded, 35.
    Red-breasted, 22, 34.
Mockingbird, 83.
Mongrel, 49.
Mother Carey's Chicken, 32.
Muddy-breast, 54.
Murre, 27.
    Brünnich's, 27.

NIGHTHAWK, 14, 65.
Nonpareil, 74.
Nuthatch, Red-breasted, 85.
    White-breasted, 16, 85.

OLD-SQUAW, 22, 39.
Oriole, Baltimore, 14, 68.
    Orchard, 14, 68.
Osprey, American, 14, 60.
Oven-bird, 15, 81.
Owl, Acadian, 63.
    American Barn, 60.
    American Hawk, 63.
    American Long-eared, 14, 61.

Owl, Arctic, 63.
    Barred, 14, 61.
    Cat, 61, 63.
    Great Gray, 62.
    Great Horned, 14, 63.
    Hoot, 61.
    Richardson's, 62.
    Saw-whet, 63.
    Screech, 14, 63.
    Short-eared, 14, 61.
    Snowy, 63.
    Sparrow, 62.
    Spectral, 62.
    White, 63.

PARTRIDGE, 56.
Peep, 50, 51.
Peet-weet, 53.
Pert, 50.
Petrel, Leach's, 32.
    Stormy, 32.
    Wilson's, 32.
Pewee, Bridge, 66.
    Wood, 14, 66.
Phalarope, Northern, 47.
    Red, 47.
    Wilson's, 48.
Phœbe, 14, 66.
Pigeon, Passenger, 88.
    Wild, 88.
Pintail, 35, 37.
Pipit, American, 83.
Plover, Black-billed, 54.
    Fool, 49.
    Golden, 54.
    Grass, 53.
    Killdeer, 54.
    Little-ring, 55.
    Piping, 13, 55.
    Ring, 55.
    Rock, 55.
    Semipalmated, 55.
    Upland, 53.
    Wilson's, 90.
Ptarmigan, Willow, 90.
Puffin, 89.

QUAIL, 55.
    American, 55.
    Marsh, 68.
Quonk-a-ree, 67.
Quwark, 44.

RAIL, Carolina, 45.
    Clapper, 89.
    King, 45.
    Red-breasted?, 45.

Rail, Virginia, 13, 45.
    Yellow, 46.
Redhead, 37.
Redpoll, 70.
    Greater, 70.
Redstart, American, 15, 83.
Red-winged, 67.
Reed Bird, 67.
Ring-neck, 55.
Robin, American, 16, 87.

SAPSUCKER, Yellow-bellied, 64.
Sanderling, 51.
Sandpiper, Baird's, 50.
    Bartramian, 53.
    Buff-breasted, 53.
    Curlew, 90.
    Least, 50.
    Pectoral, 50.
    Purple, 21, 22, 50.
    Red-backed, 50.
    Red-breasted, 50.
    Sanderling, 51.
    Semipalmated, 51.
    Solitary, 52.
    Spotted, 13, 21, 22, 53.
    Stilt, 49.
    Western Semipalmated, 51.
    White-rumped, 50.
    Wilson's, 50.
Scaup, Greater, 38.
Scoter, American, 22, 40.
    Surf, 22, 40.
    Velvet, 40.
    White-winged, 10, 11, 22, 40.
Shag, 33, 34.
"Shag, Taunton," 33, 34.
Shearwater, Cory, 31, 32.
    Greater, 32.
    Sooty, 32.
Sheldrake, 34.
    Common, 34.
    Hooded, 35.
Shitepoke, 44.
Shoveller, 36.
    Blue-billed, 38.
Sickle-bill, 53.
Shrike, Northern, 76.
    White-rumped, 76.
Siskin, Pine, 70.
Skunk-head, 40.
Smew, 35.
Snipe, Common, 49.
    English, 49.
    German, 49.
    Grass, 50.
    Horse-foot, 55.

Snipe, Red-breasted, 49.
    Robin, 49.
    Wilson's, 13, 48.
    Winter, 51.
Snowbird, 73.
Snowflake, 70.
Sora, 13, 45.
Sparrow, Chipping, 15, 73.
    English, 69.
    Field, 15, 73.
    Fox, 74.
    Grasshopper, 14, 71.
    Ground, 73.
    Henslow's, 71.
    House, 69.
    Ipswich, 71.
    Lincoln's, 73.
    Savanna, 14, 71.
    Seaside, 15, 72.
    Acadian Sharp-tailed, 72.
    Sharp-tailed, 15, 72.
    Song, 15, 73.
    Swamp, 15, 73.
    Tree, 73.
    Vesper, 14, 21, 22, 71.
    White-crowned, 72.
    White-throated, 72.
    Yellow-winged, 71.
Spoonbill, 36.
Sprig-tail, 37.
South-southerly, 39.
Swan, American, 42.
    Trumpeter, 42.
    Whistling, 42.
Swallow, Bank, 15, 76.
    Barn, 15, 75, 76.
    Chimney, 65.
    Cliff, 15, 75, 76.
    Eave, 75.
    Tree, 15, 76.
    White-breasted, 76.
Swift, Chimney, 14, 65.

TANAGER, Scarlet, 15, 75.
    Summer, 75.
Tinker, 25.
Teal, Blue-winged, 13.
    Green-winged, 36.
Tern, Black, 31.
    Caspian, 29.
    Common, 13, 20, 22, 30.
    Forster's, 29.
    Gull-billed, 89.
    Least, 30.
    Marsh, 89.
    Roseate, 22, 30.
    Sooty, 30.

Tip-up, 53.
Titmouse, Black-capped, 85.
Thistlebird, 70.
Thrasher, Brown, 15, 83.
Thrush, Alice's, 86.
    Bicknell's, 87.
    Gray-cheeked, 86.
    Hermit, 87.
    Large-billed Water, 81.
    Louisiana Water, 15, 81.
    Olive-backed, 87.
    Water, 81.
    Willow, 86.
    Wilson's, 16, 86.
    Wood, 16, 86.
Towhee, 15, 74.
"Turkey, Taunton," 33, 34.
    Wild, 88.
Turnstone, 20, 22, 55.

VEERY, 86.
Vireo, Blue-headed, 15, 77.
    Philadelphia, 77.
    Red-eyed, 15, 77.
    Solitary, 77.
    Warbling, 15, 77.
    White-eyed, 15, 77.
    Yellow-throated, 15, 77.

VULTURE, Turkey, 56.

WAKE-UP, 65.
Wamp, 39.
Warbler, Bay-breasted, 80.
    Black and White, 15, 77.
    Black and White Creeping, 77.
    Blackburnian, 80.
    Black-poll, 80.
    Black-throated Blue, 79.
    Black-throated Green, 15, 80.
    Blue-winged, 15, 78.
    Canadian, 82.
    Cape May, 79.
    Cerulean, 80.
    Chestnut-sided, 15, 80.
    Connecticut, 81.
    Golden-crowned, 79.
    Golden Swamp, 78.
    Golden-winged, 78.
    Hooded, 82.

Warbler, Kentucky, 81.
    Mourning, 82.
    Magnolia, 79.
    Myrtle, 79.
    Nashville, 15, 78.
    Northern Parula, 79.
    Orange-crowned, 78.
    Palm, 80.
    Parula, 15.
    Pine, 15, 80.
    Pine-creeping, 80.
    Prairie, 15, 81.
    Prothonotary, 78.
    Tennesee, 78.
    Wilson's, 82.
    Yellow, 15, 79.
    Yellow Palm, 81.
    Yellow-rumped, 79.
Wax-wing, Cedar, 15, 76.
Whip-poor-will, 14, 65.
Whistler, 38.
Whistler-wing, 38.
Widgeon, 35, 38.
    American, 36.
Willet, 52.
Woodcock, American, 13, 48.
Woodpecker, Downy, 14, 64.
    Golden-winged, 65.
    Hairy, 14, 64.
    Pigeon, 65.
    Red-bellied, 64.
    Red-headed, 14, 64.
    Yellow-bellied, 64.
Wren, Carolina, 15, 84.
Wren, Cotton, 79.
    Golden-crested, 86.
    House, 16, 84.
    Long-billed Marsh, 16, 85.
    Shore-billed Marsh, 16, 85.
    Winter, 84.

YELLOWBIRD, 70.
    Summer, 70, 79.
Yellow-hammer, 65.
Yellow-leg, 52.
    Bastard, 89.
    Great, 52.
    Greater, 52.
    Lesser, 52.
    Summer, 52.
    Winter, 52.
Yellow-throat, Maryland, 15, 82.

www.ingramcontent.com/pod-product-compliance
Lightning Source LLC
Chambersburg PA
CBHW031349160426
43196CB00007B/786